The Health Hazard Evaluation Program at NIOSH

Reviews of Research Programs of the National Institute for Occupational Safety and Health

Committee to Review the NIOSH Health Hazard Evaluation Program

Division on Earth and Life Studies

NATIONAL RESEARCH COUNCIL *AND*
INSTITUTE OF MEDICINE
OF THE NATIONAL ACADEMIES

THE NATIONAL ACADEMIES PRESS
Washington, D.C.
www.nap.edu

THE NATIONAL ACADEMIES PRESS • 500 Fifth Street, N.W. • Washington, DC 20001

NOTICE: The project that is the subject of this report was approved by the Governing Board of the National Research Council, whose members are drawn from the councils of the National Academy of Sciences, the National Academy of Engineering, and the Institute of Medicine. The members of the committee responsible for the report were chosen for their special competences and with regard for appropriate balance.

This study was requested by the National Institute for Occupational Safety and Health of the Centers for Disease Control and Prevention and supported by Award No. 211-2006-19152 (Task Order 001). Any opinions, findings, conclusions, or recommendations expressed in this publication are those of the author(s) and do not necessarily reflect the view of the organizations or agencies that provided support for this project. The content of this publication does not necessarily reflect the views or policies of the Department of Health and Human Services, nor does mention of trade names, commercial products, or organizations imply endorsement by the U.S. Government.

International Standard Book Number-13: 978-0-309-12647-2
International Standard Book Number-10: 0-309-12647-9

Additional copies of this report are available from the National Academies Press, 500 Fifth Street, N.W., Lockbox 285, Washington, DC 20055; (800) 624-6242 or (202) 334-3313 (in the Washington metropolitan area); Internet http://www.nap.edu.

Printed in the United States of America.

Suggested Citation: National Research Council. 2008. The Health Hazard Evaluation Program at NIOSH. Committee to Review the NIOSH Health Hazard Evaluation Program. Rpt. No. 7, Reviews of Research Programs of the National Institute for Occupational Safety and Health. Washington, DC: The National Academies Press.

THE NATIONAL ACADEMIES

Advisers to the Nation on Science, Engineering, and Medicine

The **National Academy of Sciences** is a private, nonprofit, self-perpetuating society of distinguished scholars engaged in scientific and engineering research, dedicated to the furtherance of science and technology and to their use for the general welfare. Upon the authority of the charter granted to it by the Congress in 1863, the Academy has a mandate that requires it to advise the federal government on scientific and technical matters. Dr. Ralph J. Cicerone is president of the National Academy of Sciences.

The **National Academy of Engineering** was established in 1964, under the charter of the National Academy of Sciences, as a parallel organization of outstanding engineers. It is autonomous in its administration and in the selection of its members, sharing with the National Academy of Sciences the responsibility for advising the federal government. The National Academy of Engineering also sponsors engineering programs aimed at meeting national needs, encourages education and research, and recognizes the superior achievements of engineers. Dr. Charles M. Vest is president of the National Academy of Engineering.

The **Institute of Medicine** was established in 1970 by the National Academy of Sciences to secure the services of eminent members of appropriate professions in the examination of policy matters pertaining to the health of the public. The Institute acts under the responsibility given to the National Academy of Sciences by its congressional charter to be an adviser to the federal government and, upon its own initiative, to identify issues of medical care, research, and education. Dr. Harvey V. Fineberg is president of the Institute of Medicine.

The **National Research Council** was organized by the National Academy of Sciences in 1916 to associate the broad community of science and technology with the Academy's purposes of furthering knowledge and advising the federal government. Functioning in accordance with general policies determined by the Academy, the Council has become the principal operating agency of both the National Academy of Sciences and the National Academy of Engineering in providing services to the government, the public, and the scientific and engineering communities. The Council is administered jointly by both Academies and the Institute of Medicine. Dr. Ralph J. Cicerone and Dr. Charles M. Vest are chair and vice chair, respectively, of the National Research Council.

www.national-academies.org

COMMITTEE TO REVIEW THE NIOSH HEALTH HAZARD EVALUATION PROGRAM

ROGENE F. HENDERSON, *Chair*, Lovelace Respiratory Research Institute, Albuquerque, NM
JOEL BENDER, General Motors Corporation, Detroit, MI
EULA BINGHAM, University of Cincinnati, OH
JAMES E. CONE, New York City Department of Health and Mental Hygiene
JOHN R. FROINES, Center for Occupational and Environmental Health, University of California, Los Angeles (from 07/19/2007 to 04/28/2008)
MONICA GAUGHAN, University of Georgia, Athens
CLARION JOHNSON, Exxon Mobil Corporation, Fairfax, VA
FRANKLIN E. MIRER, Hunter College of the City University of New York, New York
BARBARA SILVERSTEIN, Washington State Department of Labor and Industries, Tumwater
ROSEMARY K. SOKAS, University of Illinois, Chicago
MICHAEL J. WRIGHT, United Steelworkers, Pittsburgh, PA

National Research Council Staff

SAMMANTHA L. MAGSINO, Study Director
SUSAN R. McCUTCHEN, Senior Program Associate
TONYA E. FONG YEE, Senior Project Assistant

Preface

"Health is a state of complete physical, mental and social well-being and not merely the absence of disease or infirmity."

—Constitution of the World Health Organization

The National Institute for Occupational Safety and Health (NIOSH) Health Hazard Evaluation (HHE) Program was derived from a legislative promise to workers and employers that NIOSH would provide timely information regarding whether concentrations of substances normally found in the workplace have potentially toxic effects. NIOSH has the authority, in response to written request, to enter a workplace and measure potentially hazardous exposures, conduct medical evaluation of employees, and obtain available information from employers. Though these are essentially the same authorities provided for Occupational Safety and Health Administration (OSHA) inspections, invoked by the same parties who can trigger such inspections, health hazard evaluations are supposed to be something different. OSHA inspections are intended to ensure compliance with existing safety regulations, whereas HHEs are conducted to identify health hazards in a workplace. HHEs emphasize health effects identification through medical investigation and exposure assessment techniques combined with an epidemiological perspective, and can be at the cutting edge in identifying emerging occupational health hazards through scientific documentation of observations of those at the workplace.

At the request of NIOSH, the Committee to Review the NIOSH Health Hazard Evaluation Program was formed to evaluate the relevance, impact, and future directions of the program. The HHE Program is not a traditional research program but responds to a wide variety of requests for assistance to identify a broad spectrum of hazards. The committee's composition reflects this diversity with expertise in respiratory toxicology, public health, occupational health, industrial hygiene, risk communication, occupational medicine, occupational epidemiology, indoor environmental quality, and environmental health. The committee also has expertise in public health program evaluation (see Appendix E). It was guided, but not bound, by the evaluation criteria established by the National Academies Committee to Review NIOSH Research Programs. In this evaluation, the committee provides recommendations to improve what it believes is already a worthwhile program and provides a vision for the HHE Program of the future.

The committee thanks NIOSH staff for their kind assistance and willingness to share their knowledge. Raymond Sinclair, Teresa Schnorr, and Allison Tepper were indispensible in their efforts to answer innumerable questions from the committee. NIOSH briefers at the committee's September 2007 meeting were very helpful and included Bruce Bernard, Kay Kreiss, Teresa Schnorr, Teresa Seitz, Allison Tepper, Lewis Wade, and Ken Wallingford. Other NIOSH staff attended the meeting and contributed to the discussions, including Marlene Ackman, Fred Blosser, Christine Branche, Chad Dowell, Kelly Durst, Lynn Jenkins, Greg Lotz, Michelle Martin, Ken Martinez, John Piacentino, Doug Trout, Doug Weissman, and Ainsley Weston. Eric Landree and Valerie Williams from the RAND Corporation also spoke with the committee.

The committee received thoughtful input from a large number of stakeholders to inform its deliberations. Members particularly appreciated the willingness of several stakeholders to speak directly with the committee during its meetings. Shelley Davis (Farmworker Justice, Washington, D.C.), Janie Gittleman (Center to Protect Workers' Rights, Silver Spring, Maryland), José Olíva (Interfaith Worker Justice, Chicago, Illinois), Frank Renshaw (Rohm and Haas Company, Corydon, Pennsylvania), Marthe Kent (OSHA Region 1, Boston, Massachusetts), Andrea Kidd-Taylor (Community Health and Policy, Morgan State University, Baltimore, Maryland), Kenneth Rosenman (Occupational and Environmental Health, Michigan State University, Lansing), and Joshua Sharfstein (Baltimore City Health Department, Maryland) participated in panel discussions during the committee's second meeting. At the committee's third meeting, Linda Ayala discussed (via telecom) her experiences related to an HHE conducted for the Alameda County (California) Public Authority for In-Home Services; Peggy Hoffman, Glenn Jones, and Barbara Smisko discussed (via telecom) experiences regarding HHEs conducted at Kaiser Permanente in California; Tom Tripp and Bryant Hardy spoke with the committee (via telecom) about their HHE experiences at U.S. Magnesium in Utah; and Barbara Materna and Roger Speakman described an evaluation at a flavor and fragrance manufacturing company in California. These discussions were vital to the evaluation process and provided the committee with multiple viewpoints. The committee is also grateful to the nearly 60 respondents to its online request for information (see Appendix C). Many thoughtful comments about the HHE Program were received and provided necessary fuel for discussion. Those comments and recommendations are summarized in Appendix D.

The committee was sorry to lose the services of John Froines (Center for Occupational and Environmental Health, University of California, Los Angeles) as a member midway though the evaluation process, but it is indebted to him for his thoughtful comments on early drafts of the manuscript provided as an unpaid consultant.

This report has been reviewed in draft form by individuals chosen for their diverse perspectives and technical expertise, in accordance with procedures approved by the National Research Council's Report Review Committee. The purpose of this independent review is to provide candid and critical comments that will assist the institution in making its published report as sound as possible and to ensure that the report meets institutional standards for objectivity, evidence, and responsiveness to the study charge. The review comments and draft manuscript remain confidential to protect the integrity of the deliberative process. We thank the following individuals for their participation in the review of this report:

Susan E. Cozzens, Technology Policy and Assessment Center, Georgia Institute of Technology
James W. Curran, Rollins School of Public Health, Emory University
Edward A. Emmett, School of Medicine, University of Pennsylvania
William H. Farland, Vice President for Research, Colorado State University
Lora E. Fleming, Miller School of Medicine and Rosenstiel School of Marine and Atmospheric Sciences, Miami
Christopher C. Green, School of Medicine, Wayne State University
Dan Hair, Workers Compensation Fund of Utah
Leslie Israel, Center for Occupational and Environmental Health, University of California, Irvine
Philip J. Landrigan, Department of Community & Preventive Medicine and Children's Environmental Health Center, Mount Sinai School of Medicine
Scott Schneider, Laborers' Health and Safety Fund of North America, Washington, D.C.
Mark J. Utell, School of Medicine and Dentistry, University of Rochester
Eric Welch, Public Administration, University of Illinois at Chicago

Although the reviewers listed above have provided many constructive comments and suggestions, they were not asked to endorse the conclusions or recommendations nor did they see the final draft of the report before its release. The review of this report was overseen by **John C. Bailar,** the University of Chicago (emeritus), and **Linda Hawes Clever,** California Pacific Medical Center. Appointed by the National Research Council, they were responsible for making certain that an independent examination of this report was carried out in accordance with institutional procedures and that all review comments were carefully considered. Responsibility for the final content of this report rests entirely with the authoring committee and the institution.

Finally, the committee wishes to thank National Academies staff Sammantha Magsino, program officer; Susan McCutchen, senior program associate; and Tonya Fong Yee, senior project assistant, for the many long hours spent guiding this report through the committee process. The report could not have come to completion without their persistent efforts.

Rogene F. Henderson
Chair

Contents

Tables, Figures, and Boxes

TABLES

FIGURES

BOXES

Abbreviations and Acronyms

ACGIH	American Conference of Governmental Industrial Hygienists
AR	Authoritative Recommendations Program (of NIOSH)
ATSDR	Agency for Toxic Substances and Disease Registry
BLS	Bureau of Labor Statistics
BSC	Board of Scientific Counselors (of NIOSH)
CDC	Centers for Disease Control and Prevention
COSTEP	Commissioned Officer Student Training and Extern Program
CPWR	CPWR—The Center for Construction Research and Training (formerly the Center to Protect Workers' Rights)
DBCP	1,2-dibromo-3-chloropropane
DEOC	Director's Emergency Operations Center (of the CDC)
DRDS	Division of Respiratory Disease Studies (of NIOSH)
DSHEFS	Division of Surveillance, Hazard Evaluations, and Field Studies (of NIOSH)

EIS	Epidemic Intelligence Service
EPA	Environmental Protection Agency
Epi-X	A secure electronic communication network maintained by CDC
ERC	Education and Research Centers for Occupational Safety and Health
FACE	Fatality Assessment Control and Evaluation
FEMA	Federal Emergency Management Agency
Framework Committee	National Academies Committee for the Review of NIOSH Research Programs
Framework Document	Framework for the Review of the Research Programs of the National Institute for Occupational Safety and Health (8/10/07 version)
FSB	Field Studies Branch (of NIOSH)
FTE	Full-time equivalent
GAO	Government Accountability Office
HAZWOPER	Hazardous Waste Operations and Emergency Response Standard
HETAB	Hazard Evaluations and Technical Assistance Branch (of NIOSH)
HHE	Health hazard evaluation
HHS	Department of Health and Human Services
HP	Hypersensitivity pneumonitis
ICE	Immigration and Customs Enforcement (of the U.S. Department of Homeland Security)
IEQ	Indoor environmental quality
IMIS	Integrated Management Information System (of OSHA)
IOM	Institute of Medicine
LOHP	Labor Occupational Health Program (of the University of California, Berkeley)
MMWR	*Morbidity and Mortality Weekly Report*
MSHA	Mine Safety and Health Administration
MWF	Metalworking fluid

NAICS	North American Industry Classification System
NIEHS	National Institute of Environmental Health Sciences
NIOSH	National Institute for Occupational Safety and Health
NOPD	New Orleans Police Department
NORA	National Occupational Research Agenda (of NIOSH)
NRC	National Research Council
NTP	National Toxicology Program (of HHS)
OD	Office of the Director (of NIOSH)
OHB	Occupational Health Branch (of California)
OHSP	Occupational Health Surveillance Program (of the Massachusetts Department of Public Health)
OMB	Office of Management and Budget
OSH	Occupational safety and health
OSHA	Occupational Safety and Health Administration
OSH Act	Occupational Safety and Health Act of 1970
PEL	Permissible exposure limit
RDRP	Respiratory Disease Research Program (of NIOSH)
REL	Recommended exposure limit
RTI	Research Triangle Institute
SENSOR	Sentinel Event Notification System for Occupational Risk (of NIOSH)
SIC	Standard Industrial Classification
TLV	Threshold limit value
TOPOFF	Top Officials (a terrorism preparedness exercise)
USDA	U.S. Department of Agriculture
USPS	U.S. Postal Service
WMSD	Work-related musculoskeletal disorder
9/11	Terrorist attacks of September 11, 2001

Summary

ABSTRACT *In 2005, the National Institute for Occupational Safety and Health (NIOSH) asked the National Academies to evaluate the relevance, impact, and future directions of up to 15 of its research programs, including the Health Hazard Evaluation (HHE) Program. The HHE Program does not conduct traditional research but is mandated to respond to requests for assistance in identifying specific workplace conditions that pose health hazards to workers. In 2007, the Committee to Review the NIOSH Health Hazard Evaluation Program was formed. The committee finds the HHE Program to be highly relevant to and to have a major impact on improving occupational health, and it believes that the program should be allowed to continue to provide its services and be expanded as resources become available.*

In addition to its work to improve workplace conditions, the HHE Program responds well to emergencies, such as the terrorist attack on 9/11 and disasters related to hurricanes. Program staff are uniquely qualified to serve in leadership roles under such conditions. Training programs conducted by the HHE Program offer excellent opportunities for increasing expertise in occupational health, and the committee recommends more extensive tracking of alumni so that they can be called on for their expertise, assistance in identifying emerging issues, and emergency response.

The consensus of the committee is that the HHE Program is highly effective in investigating and advising workplaces when requested. The program fills a special need in the occupational health community by investigating unexpected or underestimated workplace hazards and relating them to worker exposures or circumstances. However, certain occupational groups might not fully benefit from program activities, particularly those from small businesses and underserved populations. The committee recommends that the HHE Program take steps to acquaint such groups with its services and elicit more requests for investigations from them. In addition, funding limitations and obligations associated with emergency response might dilute program efforts and reduce effectiveness. Finally, the committee believes that the HHE Program could help to develop a national occupational health surveillance system to facilitate recognition of emerging hazards.

On the basis of a scoring system of 1 to 5, with 5 being the highest, the committee rates both the relevance and the impact of the HHE Program as 4. If the committee had not been restricted to the use of integers, both scores would have been between 4 and 5.

OVERVIEW

In 2005, the National Institute for Occupational Safety and Health (NIOSH) asked the National Academies to evaluate the relevance, impact, and future directions of up to 15 of its research programs. One of the programs was the Health Hazard Evaluation (HHE) Program, which does not conduct traditional research but is mandated to respond to requests for assistance to identify specific workplace conditions that pose health hazards to workers. In 2007, the National Research Council formed the Committee to Review the NIOSH Health Hazard Evaluation Program.

The mission of the HHE Program is to respond to written requests to investigate potential occupational health hazards in workplaces, as defined by the Occupational Safety and Health Act of 1970 and the Federal Mine and Safety Act of 1977, and in federal agencies, including the military. The law defines who may submit requests for investigations: a request must be from an employer, a union, an employee representing at least two other employees, a single employee if the work area of concern has three or fewer employees, a federal agency health and safety committee or federal employees not covered by such a committee, or the secretary of labor (NIOSH, 2007b). Responses to requests vary from written or oral consultations on technical matters to full-scale onsite investigations. The program conducts field evaluations and consultations, responds to emergencies, and provides occupational health training for health professionals.

The committee had several discussions about how to evaluate the relevance of the HHE Program (Does the program address the right issues?) separately from its impact (Does the program address the issues effectively?). The committee evaluated the relevance of the program in terms of program inputs, activities, and outputs, as outlined in the logic model used by NIOSH to summarize the program. Program inputs evaluated include strategic planning and use of resources such as funding and staffing. Activities evaluated include the marketing of program services to all parts of the workforce, including small businesses and underserved populations, and the setting of priorities, through a triage process, for allocation of resources in response to requests for investigations. HHE Program efforts to identify emerging hazards are also evaluated. Program outputs evaluated for relevance include both formal and informal responses by the HHE Program to requests for assistance. The committee also evaluated HHE Program emergency response activities.

The impacts to be evaluated include the outcomes as listed in the HHE Program logic model, such as the reduction in worker risk and the prevention of occupational illness, the transfer of program-generated information to relevant employers and employees beyond the investigated workplaces, the influence of the HHE Program on NIOSH research and policy development, and the effect of the program on the general occupational health community, including activities of regulatory agencies, organizations of occupational safety and health professionals, and state and local health agencies.

Overall, the committee considers the program to be highly relevant and to have a major impact on improving occupational health. The program should be continued and expanded as resources become available.

RELEVANCE OF THE HEALTH HAZARD EVALUATION PROGRAM

Inputs

The strategic plan and objectives of the HHE Program are found by the committee to be highly relevant, but the committee believes that program performance measures are of necessity restricted by available resources. The committee recommends that performance measures be reviewed regularly with respect to available resources to determine whether more ambitious goals are possible.

For the most part, the HHE Program uses its resources judiciously to meet its mission in the face of a changing economy, the changing nature of HHE requests, and increased responsibilities related to emergency response. The committee recommends that the HHE Program continue to provide guidance during public health emergencies but also recommends the development of a mechanism to ensure that regular program functions continue during the deployment of key staff.

The HHE Program could make better use of available surveillance data to assist in targeting field investigations to recognize previously unknown hazards. There is no national occupational health surveillance system, but the HHE Program might influence the development of such a system, perhaps elsewhere in NIOSH, such as the Surveillance Branch of the Division of Respiratory Disease Studies.

The committee finds the extent and effectiveness of relationships between the HHE Program and federal and state health agencies to be variable and recommends the program work toward consistently effective relationships with these different groups.

Activities

One of the most important activities of the HHE Program is its response to requests for evaluation of potential occupational health hazards. The program has been a passive recipient of such requests. The committee recommends a more active approach to stimulating valid requests, especially from small businesses and underserved populations. Innovative techniques are necessary to acquaint such populations with the services offered by the HHE Program.

Another important activity is priority setting among requests for investigations. For that purpose, the HHE Program has developed a triage process to determine whether a request meets regulatory requirements and whether a site visit and full investigation are warranted. The committee finds the development of this process a program improvement, but notes that the process is neither well documented nor transparent to the requestor. The committee recommends the development of an explicit, transparent written process for triage of requests.

Outputs

The major outputs of the HHE Program are reports written in response to requests. The committee finds that the reports are generally well written, present relevant information supported by appropriate documentation, and reflect a high level of expertise. However, the committee recommends the development of a well-defined quality assurance program that incorporates expert review from elsewhere in NIOSH and externally from the professional and research communities. Such a program could ensure consistently high quality outputs.

An outcome of the HHE Program should be the detection of emerging hazards based on requests received. The committee finds no systematic approach to achieving such an output and recommends the HHE Program initiate a formal periodic assessment of new and emerging hazards.

IMPACT OF THE HEALTH HAZARD EVALUATION PROGRAM

Reducing Worker Risk and Preventing Occupational Illness

In the field of occupational health, evidence suggests that exposure to certain chemical or physical agents causes illness, injury, or death. Elimination of exposure can reduce the number of cases of disease in those already exposed, prevent disease in new hires, or both. On the basis of the cause-effect relationship, attempts are made to reduce disease risk by reducing or eliminating exposure to specific agents. Numbers of cases of disease or numbers of deaths avoided can be estimated on the basis of reduction in exposure, but ideally one would have quantitative evidence of the reduction in illness or death at specific investigated worksites or similar workplaces throughout the country. Such quantitative evidence of impact is generally unavailable.

Transferring Program-Generated Information to Relevant Employers and Employees Beyond Investigated Workplaces

The HHE Program uses a wide variety of mechanisms to disseminate information, including information posted on the NIOSH website; free CDs made available to industries, workers, professors, and students; published articles in technical, trade, and scientific journals; and when required, Health Hazard Alerts to other government agencies at both state and federal levels, such as the Occupational Safety and Health Administration (OSHA). The committee finds, however, that the penetration of this information into some communities is variable, with less penetration into small businesses unaffiliated with trade organizations and underserved populations, including migrant workers. The committee recommends the development and use of innovative techniques to reach such populations.

Influence of the Health Hazard Evaluation Program on NIOSH Research and Policy Development

The HHE Program does not have the authority to promulgate regulations, but it does inform and support NIOSH-recommended guidelines and policies and NIOSH testimony regarding proposed OSHA rules. These activities have included guidance on a variety of occupational hazards, including those associated with tuberculosis, ergonomics, biosolids, latex, indoor air, metalworking fluids, histoplasmosis, hexavalent chromium, body art, hearing loss, and respirator selection decision logic. Thus, the HHE Program has had an impact on a broad array of issues related to development of NIOSH policy to reduce worker risk associated with occupational hazards. As stated earlier, quantitative data are not available to determine exact reductions in illness and death (the end outcomes), but there can be no doubt that the HHE Program is a major contributor to the development of policies expected to lead to a safer working environment (an intermediate outcome).

The HHE Program has also had both direct and indirect impacts on NIOSH research programs as well as the body of scientific knowledge in general. HHE Program reports influence the direction taken by the NIOSH Respiratory Diseases Research Program through the identification of unexpected workplace hazards. Examples of research influenced by the HHE Program are studies of the respiratory problems caused by flavorings, flock,[1] waterproofing spray, vaporized viruses, and more recently, nanoparticles. Extensive publications of those studies can be found in the open literature. HHE Program staff indicated to the committee that its reports also influence research in other NIOSH programs, such as the Cancer, Reproductive, and Cardiovascular Diseases Program; the Engineering Controls Program; the Exposure Assessment Program; the Hearing Loss Prevention Program; the Musculoskeletal Disorders Program; and the Personal Protective Technology Program.

Important additional impacts of the HHE Program result from the development of occupational expertise through its training programs. These are major contributions, and program impact could be expanded by increasing recruitment of trainees and by maintaining contact with training program alumni. Training program participants and alumni could be enlisted to provide expertise, assistance in identifying emerging hazards, and assistance in maintaining routine program operations during emergency response.

Transfer of Program-Generated Hazard and Prevention Information to the Occupational Health Community

Program-generated hazard and prevention information is transferred in part by HHE Program reports and technical assistance letters generated in response to requests for assistance. In addition, HHE Program staff publish in the peer-reviewed literature, present research at major academic conferences related to occupational health, publish in trade journals, and make presentations at trade association meetings. HHE Program staff have developed compendia of findings and recommendations about well-understood occupational health risks. Completed compendia include those on isocyanates, noise, tuberculosis, and lead (NIOSH 2004e, 1998c, 2001b, and 2001a, respectively). The committee encourages the program to continue the development of such compendia as a valuable method

[1]Fine, small-diameter synthetic fibers—the breathing of which can cause respiratory problems.

of transferring information to the general occupational health community. The program also participates in a variety of interagency activities related to occupational health.

The HHE Program is a recognized federal leader in responding quickly and effectively to new and emerging hazards, particularly in emergencies. A challenge for the program is continuing routine operations while responding to emergencies.

Despite the activities described above, the committee believes that more could be done to reach a broader variety of people, trades, and businesses, such as small, high-risk residential construction companies, small businesses, and immigrant or temporary workers. The committee heard from some state and local health officials who were unaware of the existence of the HHE Program. The program should not be a secret to those interested in the safety of the workplace.

Evaluation of the impact of the HHE Program would be greatly enhanced by the establishment of a disciplined, after-the-fact evaluation of the relevance, impact, and quality of responses to HHE investigations. A "followback" program was initiated on the advice of an earlier evaluation committee, but it should be expanded to include more complete information on the impact of the program.

FUTURE DIRECTIONS OF THE HEALTH HAZARD EVALUATION PROGRAM

The committee's "vision" for the future of the HHE Program is that the program would serve to identify heretofore unrecognized workplace hazards, as well as known hazards for which permissible exposure limits or other control measures appear inadequate. The HHE Program of the future would continue to do what NIOSH can do uniquely—for example, emphasizing health effects identification that combines medical investigation and industrial hygiene (exposure assessment) techniques with epidemiological and toxicological perspectives, which may lead to the development of new measurement and control techniques.

To achieve such goals, the HHE Program must become better known in the workplace and in the occupational health community. The program can only respond to requests from others, but such requests cannot be generated if potential requestors have not heard of the program. A major challenge will be to make its services more widely known to small businesses, underserved populations, and others.

In the committee's vision, the HHE Program of the future would expand its role as a major source of training of occupational health professionals, not only for NIOSH, but for other government agencies, academic institutions, and international groups as well. The program would maintain good working relationships with local, state, and national organizations with ties to occupational health so that it would be a primary resource when health problems arise in the workplace. The HHE Program of the future would make full use of its recently instituted followback program and maintain records of the effectiveness of recommended workplace interventions. In that way, HHE Program staff could evaluate how well the program meets its strategic goals and learn which techniques are most effective. Finally, the HHE Program of the future would continue to provide guidance and recommendations during public health emergencies.

In contrast, the HHE Program of the future would not expend its limited resources doing what others can do, such as conducting routine industrial hygiene measurements or providing standardized responses to common problems. The expertise in the HHE Program is better used in developing generalized protocols for responding to common problems, rather than investigating routine industrial hygiene matters. HHE Program investigations must also be distinguished from OSHA compliance inspections and small-business consultations. The HHE Program is designed to address emerging hazards, not to supplement the OSHA enforcement process or small-business consultation program.

The committee's recommendations are summarized in Box S-1.

SCORING

The scoring system developed by the Committee to Review NIOSH Research Programs is based on a five-point scale in which 5 is the highest score (see Box S-2). The committee believes that the HHE Program scores somewhere above a 4 for both impact and relevance, but notes where the program could be improved. Because the committee was required to provide integer scores, the HHE Program received a 4 for both relevance and impact.

BOX S-1
Committee Recommendations

The committee organizes its recommendations in eight categories, presented in an order consistent with the HHE Program logic model. The committee does not set priorities among its recommendations; it prefers encouraging implementation of any of them as resources are available. The committee recognizes that implementation of many of these recommendations will depend on the availability of resources, but at the request of NIOSH, it refrains from making recommendations regarding resource allocations. Therefore, general recommendations are made for program improvement. The next step for the HHE Program is to consider how resources could be allocated to allow implementation.

Recommendation 1: Conduct regular assessments of performance measures to determine whether available resources allow more ambitious goals.

Recommendation 2: Improve the mechanisms by which requests for HHEs are sought and prioritized to include a broader array of requests from a wider variety of requestors. The program could achieve this through

a. Systematic use of professional meetings, scientific conferences, scientific literature, and surveillance data, including those generated by NIOSH, to assist in prioritizing field investigations and in recognizing emerging issues.
b. Implementing, as part of the triage process, a formal technical assistance mechanism to help requestors to formulate valid HHE requests. In cases where an HHE is not appropriate or where resource limitations prohibit an investigation, technical assistance should include referral to more appropriate NIOSH divisions or government agencies.
c. Development of an explicit, written process for classifying and prioritizing HHE requests.
d. Better formalizing of the triage process, including the identification of needed expertise, and improving the transparency of the process to HHE requestors, while maintaining flexibility and speed.
e. Establishment of formal relationships with organizations representing underserved populations, small businesses, and their employees.
f. Enhancing HHE Program outreach to OSHA national and regional offices and to state health and labor departments to better communicate the functions and activities of the HHE Program, increase cooperation with these agencies, and provide more complete and timely feedback.

Recommendation 3: Ensure that recommendations in HHEs are relevant, feasible, effective, and clearly explained. Such steps may include

a. Explanation of the relevance, feasibility, and impact of each recommendation in the text of HHE reports.
b. Priority-setting among recommendations in all reports to indicate those requiring immediate action in the targeted workplace.
c. Debriefing in NIOSH after site visits and report dissemination for determination of relevance and impact on a systematic basis (potentially missed opportunities to identify emerging health hazards could also be identified).

continued

BOX S-1
Continued

d. Modification of the followback surveys for use in assessing the relevance, feasibility, and impact of recommendations.

e. Enhancement of internal quality assurance by development of a formal program that may include the external review of a sampling of recent reports and technical assistance letters for scientific content, report completeness, and appropriateness of recommendations.

Recommendation 4: Use the HHE Program to develop occupational health professional resources. This could be accomplished through

a. Increased recruitment of new investigators from universities, the Epidemic Intelligence Service (EIS), the Commissioned Officer Student Training Extern Program (COSTEP), occupational medicine residencies, Education and Research Centers for Occupational Safety and Health (ERCs), and state and local health departments into HHE Program training rotations.

b. Tracking and mobilizing the extensive talent and commitment represented in the HHE Program-trained occupational health workforce. A network of HHE Program alumni could be fostered to help develop HHE opportunities. A program-level advisory board could assist the program in leveraging resources, serve a recruiting and retention function, assist in identifying emerging issues, and provide expert advice during normal program operations and when normal program operations are interrupted by emergency response activities.

c. Engagement and use of ERCs and other university-based training programs to involve trainees in HHE field investigations.

d. More formal collaboration with ERC faculty and other extramural researchers to assist in field investigation, dissemination, and training opportunities.

Recommendation 5: Develop a proactive, comprehensive information-transfer strategy for HHE Program outputs with better approaches to reaching wider audiences, including traditionally underserved populations. The HHE Program could

a. Use innovative techniques to reach small businesses and underserved populations, creating a broad array of mechanisms for communicating with diverse constituencies and attending to issues of literacy, language, and national-origin barriers. The effectiveness of applied outreach should be evaluated in a formal manner.

b. Improve the searchability of the online HHE search engine by developing a list of standardized key words (an alphabetized list of hazards and diseases would be beneficial).

c. Develop distribution mechanisms that are not Internet-dependent to complement Internet distributions.

d. Disseminate HHE results more broadly to groups likely to be affected, including distribution of HHE reports in the geographic regions where investigations are conducted.

e. Increase efforts to compile compendia of findings (such as those developed for isocyanates, noise, tuberculosis, and lead) when generalized process-oriented findings can be gleaned from the experience of the HHE Program in a variety of workplace settings.

f. Develop improved methods of outreach to stakeholders so that workers and workplaces affected by new and emerging occupational health problems will be alerted quickly.

g. Supplement program outreach efforts by using community and small-business groups to translate HHE results and findings for their constituencies.
h. Leverage existing NIOSH, Centers for Disease Control and Prevention (CDC), and Department of Health and Human Services (HHS) resources to enhance technology transfer.
i. Evaluate, in a formal manner, the effectiveness of information-transfer programs, including knowledge transfer to employers and employees at worksites where HHEs have not been conducted.

Recommendation 6: Develop more extensive formal linkages and mechanisms with other parts of NIOSH, CDC, and HHS to enhance the capacity for involvement in policy-relevant impacts through
a. Promotion and increase in direct communication, especially with OSHA and state occupational safety and health agencies.
b. Alerts to NIOSH and CDC about HHEs that are relevant to policy-making outside the CDC system.
c. Continued regular use of the National Occupational Research Agenda (NORA) sector councils and the NIOSH Board of Scientific Counselors to disseminate information about the HHE Program.
d. Pursuit of a change in the HHE Program's legislative and regulatory authority to improve the capacity to identify hazards in need of HHEs, improve the ability to gain entrance to facilities when requested by treating physicians or community representatives, and address exposures other than chemical agents.

Recommendation 7: Initiate formal periodic assessment of new and emerging hazards. To accomplish this, the HHE Program could
a. Evolve from a program that passively receives requests to a proactive program that seeks opportunities for field investigations.
b. Develop systematic approaches to identify hazards where OSHA permissible exposure limits are inadequate or nonexistent, to identify unknown hazards, and to identify known hazards encountered under new circumstances.
c. Establish and periodically review a tickler file of inconclusive or unexpected evaluation results to determine whether new trends or problems may be emerging.
d. Periodically meet with intramural and extramural research scientists and stakeholders in government, academe, labor, and industry to discuss specific unresolved evaluations, to review aggregate findings, and to solicit input about new or emerging hazards or interventions.

Recommendation 8: Continue to provide guidance and recommendations during public health emergencies. To accomplish this, the HHE Program could
a. Remain diligent by working with NIOSH management to avoid negative impact on routine activities of the HHE Program as a result of emergency response activities.
b. Develop a mechanism, such as the enlistment of help from training program participants and alumni, to ensure continuation of routine operations in the absence of staff involved in emergency response.

BOX S-2
Committee Scoring Criteria

Relevance

5 = Activities are in high-priority subject areas and NIOSH is significantly engaged in appropriate transfer activities for completed projects/reported results.
4 = Activities are in priority subject areas and NIOSH is engaged in appropriate transfer activities for completed projects/reported results.
3 = Activities are in high-priority or priority subject areas, but NIOSH is not engaged in appropriate transfer activities; or activities focus on lesser priorities but NIOSH is engaged in appropriate transfer activities.
2 = Activities are focused on lesser priorities and NIOSH is not engaged in or planning some appropriate transfer activities.
1 = Activities are not focused on priorities and NIOSH is not engaged in transfer activities.

Impact

5 = Activities have made major contribution(s) to worker health and safety on the basis of end outcomes or well-accepted intermediate outcomes.
4 = Activities have made some contributions to end outcomes or well-accepted intermediate outcomes.
3 = Activities are ongoing and outputs are produced that are likely to result in improvements in worker health and safety (with explanation of why not rated higher). Well-accepted outcomes have not been recorded.
2 = Activities are ongoing and outputs are produced that may result in new knowledge or technology, but only limited application is expected. Well-accepted outcomes have not been recorded.
1 = Activities and outputs do not result in or are NOT likely to have any application.

NA = Impact cannot be assessed; program not mature enough.

1

Introduction

CHARGE TO THE COMMITTEE

In 2005, the National Institute for Occupational Safety and Health (NIOSH) requested that the National Academies review up to 15 of its research programs. The Committee to Review the NIOSH Health Hazard Evaluation (HHE) Program was formed in 2007 to carry out an independent evaluation of the HHE Program under the following charge:

> The National Academies will appoint an ad hoc committee to carry out an independent evaluation of the impact, relevance, and future directions of the NIOSH Health Hazard Evaluation (HHE) Program. The committee will evaluate not only what the NIOSH HHE program is producing, but will also determine whether it is appropriate to credit NIOSH activities with changes in workplace practices, hazardous exposures, and/or occupational illnesses and injuries, or whether the changes are the result of other factors unrelated to NIOSH.
>
> In conducting its assessment, the committee will evaluate
>
> 1. The impact of the program on
>   - reducing worker risk and preventing occupational illness in investigated workplaces;

- transferring program-generated information to relevant employers and employees beyond the investigated workplaces;
- NIOSH research and policy-development programs; and
- the activities of regulatory agencies, occupational safety and health professionals and organizations, state and local health agencies, and others in the occupational health community, as achieved by transferring program-generated hazard and prevention information.

2. The relevance of the program in addressing current and emerging workplace health hazards.

   The evaluation committee will provide quantitative assessments in the form of integer scores (on a scale of 1-5) for both the relevance and impact of the program in addressing workplace health hazards, to be accompanied by qualitative assessments of all the categories above. The committee will develop its own methodology for the evaluation, guided by the methodology and framework developed by the Committee to Review NIOSH Research Programs where appropriate.

The HHE Program does not conduct traditional research as do other NIOSH programs. It is mandated to respond to requests for assistance to identify specific workplace conditions that pose health hazards to workers (NIOSH, 2007b). The charge to this committee was therefore modified from that given for other evaluations in this series.

Definitions of Impact and Relevance

In developing its evaluation methodology, the committee carefully studied the charge and determined definitions for impact and relevance. For this evaluation, impact is defined more broadly than for the evaluations of other NIOSH programs. This committee is asked to determine whether the HHE Program contributes to occupational health in the workplace, as well as whether the program positively affects policy, other research programs, and the health and safety community in general. The charge to the committee specifically includes transfer of information as part of impact, although the methodology and evaluation framework (hereafter called the Framework Document) developed by the Committee to Review NIOSH Research Programs (hereafter called the Framework Committee) tends to include transfer in its definition of relevance. The committee considers transfer activities important for both relevance (Is the program doing the right things?) and impact (Is what the program is doing effective?).

The Framework Document defines a relevant program as an integrated program involving interrelated surveillance,[1] research, and transfer activities. The evaluation criteria include the severity or frequency of hazards addressed and the number of people at risk; the extent to which gender-related issues and those of vulnerable populations (hereafter described as "underserved populations"; see Box 1-1 for definition) are addressed; the extent to which the health and safety needs of small businesses are addressed; the "life stage" of problems being addressed;[2] and the structure and content of the program.

The criteria for relevance as defined in the Framework Document are as applicable to the HHE Program as they are to traditional NIOSH research programs. The committee is to determine whether occupational health issues addressed by the HHE Program are the most serious in gravity or the most frequent among the spectrum of issues to which resources might be applied. It is also to determine whether the HHE Program addresses the needs of underserved populations and small businesses. HHEs are not the same as investigator-initiated research such as that sponsored by the National Institutes of Health. HHEs are considered part of the federal government's public health surveillance activities, similar to outbreak investigations of infectious diseases in communities, and thus are exempt from the requirement of review by institutional review boards to which other forms of human subjects research are subject. For this reason, the committee considers the extent to which HHEs respond to the needs of the requesters to be an additional measure of relevance.

In all its criteria, the Framework Document directs evaluation committees to consider how program research is relevant or has impact. This committee finds that substituting program activities for "program research" is an adequate modification in most cases, including within the Framework Committee's scoring criteria. Box 1-2 provides the scoring criteria used by the committee in the determination of rankings for relevance and impact.

---

[1]Public health surveillance is the "ongoing, systematic collection, analysis, interpretation, and dissemination of data regarding a health-related event for use in public health action to reduce morbidity and mortality and to improve health. Data disseminated by a public health surveillance system can be used for immediate public health action, program planning and evaluation, and formulating research hypotheses" (MMWR, 2001b:2).

[2]The life stage of an issue is determined by identifying the type of activities that should be undertaken at a given time to correct an issue. For example, as health effects are understood, efforts should shift from efficacy to intervention and intervention-effectiveness research.

BOX 1-1
Vulnerable or Underserved Populations

Vulnerable populations are defined in the Framework Document (Appendix A) as groups of workers who have biologic, social, or economic characteristics that place them at increased risk for work-related conditions or on whom inadequate data have been collected. They include disadvantaged minorities, disabled persons, low-wage workers, and non-English speakers for whom language or other barriers present health or safety risks. Undocumented workers also fall within this category. This evaluation committee uses the term "underserved populations" when referring to vulnerable populations.

THE EVALUATION PROCESS

Information Gathering

The committee reviewed material submitted by the HHE Program in the form of an "evidence package" that describes the HHE Program's legislative foundations, organizational structure and management, resources, activities, and outputs (NIOSH, 2007b). Contributions to improvements in worker health are summarized in the form of narratives related to specific workplace issues. Results of previous evaluations, and lists and examples of HHE Program reports are provided. HHE Program staff also responded during meetings and in writing to numerous questions from the committee and participated in a telephone conference with committee members to discuss HHE Program emergency response activities. Appendix B is a list of materials provided by the HHE Program to assist the evaluation process.

To deliberate its charge, the committee met four times. Two meetings were held in Washington, D.C. (September and December 2007), and two in Irvine, California (January and February 2008). The first three meetings included briefings from NIOSH and HHE Program staff and stakeholders. Meeting agendas, including names and affiliations of those who provided input, are given in Appendix C. Committee members discussed pertinent issues with several stakeholders on two panels representing the workplace and public sectors during the December 2007 meeting. In January 2008, the committee heard from stakeholders regarding four specific HHEs (NIOSH, 2004a, 2005b, 2005f, 2007i). Meetings were held on both the East and the West Coast to facilitate stakeholder participation and attendance. Subsets of the committee held numerous phone conferences and a meeting in Cincinnati, Ohio, to prepare the final report.

BOX 1-2
Scoring Criteria

The following is the scoring criteria used by the committee in its evaluation of relevance and impact of the HHE Program. The criteria are based on those developed by the Framework Committee (see Appendix A), but have been slightly modified to accommodate the more applied nature of the HHE Program.

Scoring Criteria for Relevance

5 = Activities are in high-priority subject areas and NIOSH is significantly engaged in appropriate transfer activities for completed projects/reported results.
4 = Activities are in priority subject areas and NIOSH is engaged in appropriate transfer activities for completed projects/reported results.
3 = Activities are in high-priority or priority subject areas, but NIOSH is not engaged in appropriate transfer activities; or activities focus on lesser priorities but NIOSH is engaged in appropriate transfer activities.
2 = Activities are focused on lesser priorities and NIOSH is not engaged in or planning some appropriate transfer activities.
1 = Activities are not focused on priorities and NIOSH is not engaged in transfer activities.

Scoring Criteria for Impact

5 = Activities have made major contribution(s) to worker health and safety on the basis of end outcomes or well-accepted intermediate outcomes.
4 = Activities have made some contributions to end outcomes or well-accepted intermediate outcomes.
3 = Activities are ongoing and outputs are produced that are likely to result in improvements in worker health and safety (with explanation of why not rated higher). Well-accepted outcomes have not been recorded.
2 = Activities are ongoing and outputs are produced that may result in new knowledge or technology, but only limited application is expected. Well-accepted outcomes have not been recorded.
1 = Activities and outputs do not result in or are NOT likely to have any application.

NA = Impact cannot be assessed; program not mature enough.

The committee found a general paucity of quantitative data showing the impact of HHE Program activities. For this reason, the committee made great efforts to hear from those who could reasonably be expected to benefit from HHE Program activities. The committee invited individuals representing small and large businesses; researchers in academe; local, state, and federal agencies; labor unions; and workers' rights organizations and advocacy groups to participate in discussions regarding the relevance and impact of HHE Program activities. All were asked to speak of their positive and negative experiences with the HHE Program and the usefulness and effectiveness of any resulting HHE recommendations. HHE Program stakeholders unable to travel to meetings were invited to participate in discussions via teleconference. The committee found it necessary to rely on this anecdotal evidence when no other evidence was available, but it is careful to provide citations in the report when doing so.

To receive a broader range of stakeholder input, the committee also requested public input via an online questionnaire (see Appendix C). The questionnaire was announced via e-mail to more than 500 stakeholders. Responses could be submitted online, by e-mail, or by standard mail, and the option to respond anonymously was available. Questionnaire announcements were distributed to individuals suggested by the committee, as well as to those on lists provided by the HHE Program and generated by other committees in this evaluation series. The HHE Program provided links to the questionnaire on its Internet site, and the request for input was also announced through a variety of means including a NIOSH listserv for state health and labor departments that partner with NIOSH on occupational health surveillance; a listserv for occupational health practitioners run by the University of North Carolina; the American Industrial Hygiene Association weekly e-newsletter; and NIOSH eNews, a monthly newsletter (NIOSH eNews, 2008). The committee received 57 responses. A summary of stakeholder recommendations and identified emerging issues is provided in Appendix D (derived from Stakeholder Response Table, 2008). The committee found stakeholder insights thoughtful and invaluable during deliberations.

Additional stakeholder input related to HHE Program emergency response activities was solicited (see Appendix C). Two responses were received as a result of that effort.

Period Evaluated

Because the HHE Program has evolved since its inception and because there have been multiple evaluations of this program in the past (see Chapter 2), the committee focuses on current program processes and activities. The committee chooses to emphasize the period between 1997 and 2007 in its evaluation; however many program activities before and after this time frame are considered.

REPORT ORGANIZATION

This report is organized into five chapters. This chapter orients the reader by outlining the committee review process. Chapter 2 provides a short history of the HHE Program and its legislative authority and a description of the program's inputs, activities, and outputs. The committee's evaluation of the HHE Program can be found in Chapters 3 and 4, which are organized in a manner consistent with the program's logic model (presented in Chapter 2). In Chapter 3, the committee describes the relevance of the program's strategic plan, its use of resources, program activities, outputs, and emergency response activities. The committee's findings are organized and described under each of these categories. In Chapter 4, the committee evaluates the impact of the program. The chapter includes four sections corresponding to the four parts of the committee's charge with respect to impact: (1) reduction in risk and prevention of occupational illness in investigated workplaces; (2) transfer of information beyond investigated worksites; (3) influence on NIOSH research and policy; and (4) impact on others in the occupational health community. A fifth section discusses the impact of program emergency response activities. Finally, in Chapter 5, the committee describes its vision of what the HHE Program of the future would be and summarizes recommendations made in Chapters 3 and 4.

2

Description of the Health Hazard Evaluation Program

INTRODUCTION

The primary charge to the Health Hazard Evaluation (HHE) Program, as described in Section 20(a)(6) of the 1970 Occupational Safety and Health Act (OSH Act; P.L. 91-596),[1] is to respond to written requests to investigate workplace health hazards. The program conducts field evaluations and consultations, responds to emergencies, and provides training for health professionals. These activities are conducted by staff trained as generalists in occupational medicine, epidemiology, and industrial hygiene, and by some specialists in areas such as pulmonary medicine (see Box 2-1 for definitions of some of these terms). HHE Program stakeholders include those at worksites where HHEs are conducted, workers and employers at similar workplaces, regulatory bodies, occupational and public health practitioners, other National Institute for Occupational Safety and Health (NIOSH) programs, and the research community more broadly.

The Committee to Review the HHE Program opted to separate the components to be evaluated as described in the HHE Program logic model (Figure 2-1). To be evaluated are the program's inputs (strategic goals and objectives, program resources, and communication from stakeholders), activities (responses to HHE requests), outputs (such as HHE reports and NIOSH published documents),

[1]P.L. 91-596, 91st Cong., S.2193, December 29, 1970, as amended through January 1, 2004 (http://www.osha.gov/pls/oshaweb/owadisp.show_document?p_table=OSHACT&p_id=2743, accessed March 21, 2008).

BOX 2-1
Definitions of Common Terms

Epidemiology The study of the distribution of, and the physical, biological, social, cultural, and behavioral factors that influence, health-related states or events in specified populations. Epidemiology also includes the application of this study to control of health problems (SOURCE: Last, 2001).

Industrial hygiene The science and art devoted to the recognition, evaluation, and control of the environmental factors or stresses arising from or in the workplace that may cause sickness, impaired health and well-being, or significant discomfort and inefficiency among workers or among persons in the community; the profession that anticipates and controls unhealthy conditions of work to prevent illness among employees (SOURCE: Last, 2001).

Occupational medicine The specialized practice of medicine, public health, and ancillary health professions in an occupational setting in order to promote health and prevent occupationally related disease and injury (SOURCE: Last, 2001).

Surveillance The systematic collection, analysis, interpretation, and dissemination of data regarding a health-related event for use in public health action to reduce morbidity and mortality and to improve health (SOURCE: MMWR, 2001b:2).

intermediate outcomes (such as reductions in hazardous exposures), and end outcomes (reduction in occupational illness). A simplified version of the logic model (shown in Appendix A, Figure 2) has been used in reviews of the other NIOSH research programs.

This chapter provides a historical context for the program and describes HHE Program inputs, activities, and outputs. The relevance of these components is assessed in Chapter 3. Stakeholder response to and impacts resulting from HHE activities are discussed primarily in Chapter 4.

HISTORICAL OVERVIEW

The Occupational Safety and Health Administration (OSHA) is the federal agency charged with protecting worker safety and health by setting and enforcing workplace standards. However, many hazards, such as certain chemicals used in commerce or conditions that cause musculoskeletal disorders are not regulated by standards, and many existing standards may be obsolete. When occupational illness is suspected, OSHA workers and employers rely on the HHE Program to identify illness-causing hazards and to recommend control solutions.

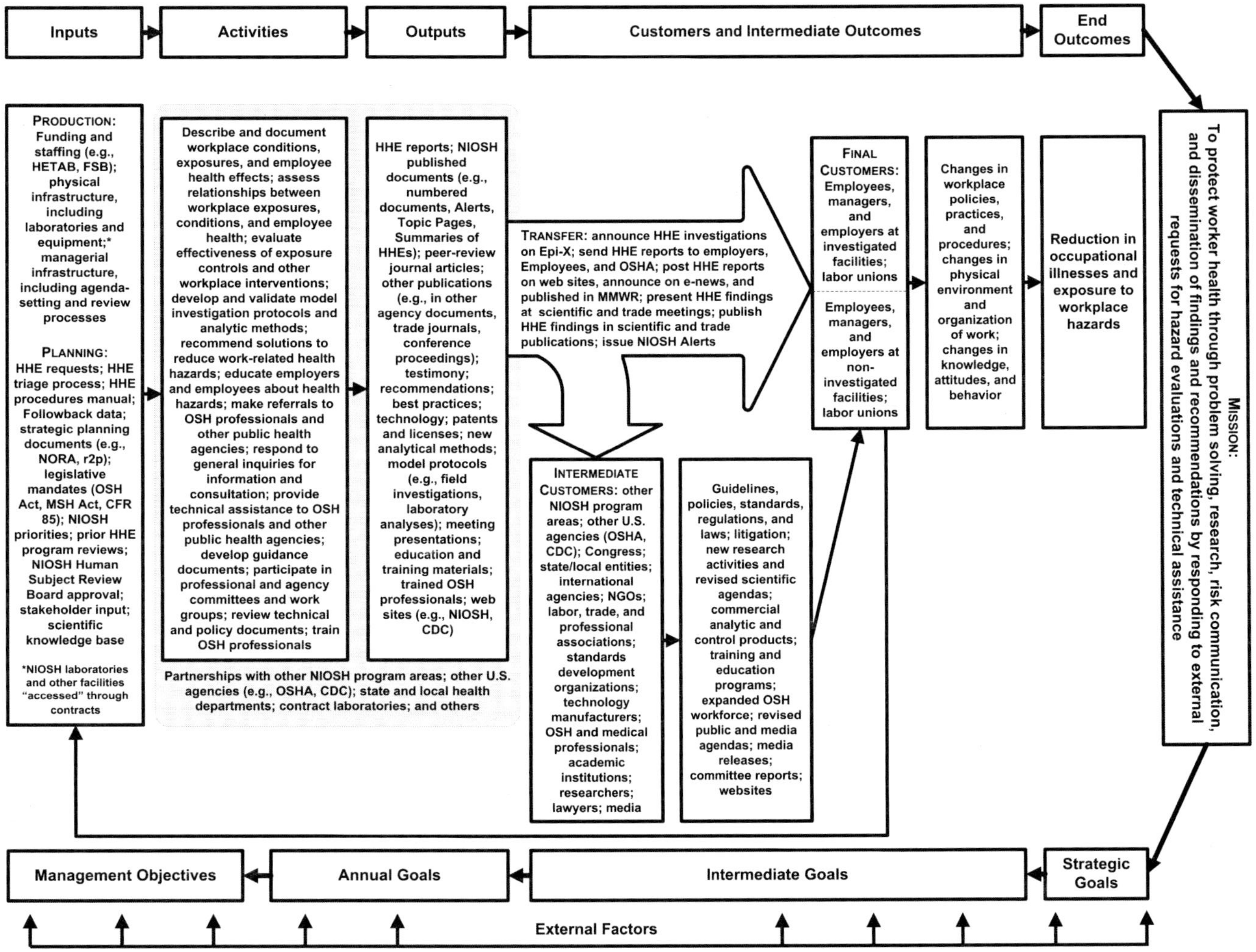

FIGURE 2-1 Logic model of the HHE Program.
SOURCE: NIOSH (2007b).

During the June 19, 1968, Senate hearings on the proposed OSH Act, Phillip R. Lee, M.D., Assistant Secretary for Health and Scientific Affairs, Department of Health, Education, and Welfare, testified:

> We are dealing with hidden health hazards. Many occupational illnesses occur only after long periods of exposure to one or more hazards in the environment so that the link between hazard and frank disease is not dramatic and overt. . . . Moreover, a large part of the problem lies in small establishments which employ fewer than 100 workers. . . . The vast majority of these workplaces have no safety engineers, doctors, nurses, hygienists, or meaningful safety and health programs. . . . The villain is not greed; it is ignorance. One of the basic objectives of the legislation before you today . . . is to remove those barriers of ignorance that result in so much needless suffering and economic loss (U.S. Congress, Senate, Committee on Labor and Public Welfare, 1968:270-271).

Congress ultimately recognized that the nation needed a mechanism to investigate health hazards in the workplace, regardless of compliance with specific standards. The OSH Act of 1970 (P.L. 91-596) included provisions in Section 20 (Research and Related Activities, 29 USC 669)[2] to be carried out by the Secretary of Health, Education and Welfare through a new NIOSH, which was established in Section 22 of the act.[3]

Section 20(a)(6) provides that

> The Secretary of Health and Human Services shall publish within six months of enactment of this Act and thereafter as needed but at least annually a list of all known toxic substances by generic family or other useful grouping, and the concentrations at which such toxicity is known to occur. He shall determine *following a written request by any employer or authorized representative of employees, specifying with reasonable particularity the grounds on which the request is made, whether any substance normally found in the place of employment has potentially toxic effects in such concentrations as used or found* [emphasis added]; and shall submit such determination both to employers and affected employees as soon as possible. If the Secretary of Health and Human Services determines that any substance is potentially toxic at the concentrations in which it is used or found in a place of employment, and such substance is not covered by an occupational safety

[2]http://www.osha.gov/pls/oshaweb/owadisp.show_document?p_table=OSHACT&p_id=3374 (accessed March 21, 2008).

[3]29 USC 671 (http://www.osha.gov/pls/oshaweb/owadisp.show_document?p_table =OSHACT&p_id =3376 (accessed March 21, 2008).

> or health standard promulgated under section 6, the Secretary of Health and Human Services shall immediately submit such determination to the Secretary [of Labor], together with all pertinent criteria.

The organization that conducts these investigations became the HHE Program, which responds to requests from employers, employees, and their representatives, and other agencies[4] to conduct workplace evaluations. Through HHEs, NIOSH identifies current health hazards and makes recommendations to reduce exposures and prevent disease and disability. HHEs may be conducted in any private, federal, or other government workplace. Authority is governed by laws (P.L. 91-596,[5] P.L. 95-164[6]), federal regulations (29 CFR 1960,[7] 42 CFR 85[8]), and Executive Order 12196 of February 26, 1980. The language of the law is important because it implies that the HHE Program should address hazards not heretofore recognized or levels of exposure not previously deemed hazardous. The present committee concludes that Congress intended to fill data gaps present in 1970, when OSHA adopted wholesale as permissible exposure limits (PELs) (Robinson, 1991) many of the existing American Conference of Governmental Industrial Hygienists (ACGIH) threshold limit values (TLVs) (ACGIH, 1989). NIOSH was given authority to enter the workplace following a written request, and legislation required certain employers to cooperate with an investigation by allowing exposure measurements, making employees available for medical evaluation, and providing available information. While right of access resembles that for an OSHA inspection, the HHE Program conducts evaluations rather than inspections. There are important differences. OSHA inspects for known hazards and seeks to enforce safety and health regulation by issuing citations and fines as appropriate. By contrast, the HHE Program has not been given regulatory authority; its mission is intended to evaluate unknown situations with the goal of identifying previously unrecognized hazards and developing new control strategies.

Who may submit requests for investigations is defined by law: the request must be from an employer; a union; an employee representing at least two other employees; a single employee if the work area of concern has three or fewer employees;

[4]29 CFR 1960, Basic Program Elements for Federal Employees Occupational Safety and Health Programs and Related Matters. Section 1960.35 of these regulations describes the procedures for requesting HHEs in federal agency workplaces. NIOSH follows the procedures outlined in the regulations governing HHEs (42 CFR 85) when evaluating federal agency workplaces (http://www.osha.gov/pls/oshaweb/owadisp.show_document?p_table=STANDARDS&p_id=11284, accessed October 10, 2008).

[5]29 USC 667 (Section 18(c)(6)), 29 USC 668 (Section 19), and 29 USC 669 (Section 20(a)(6)).

[6]Section 501(a).

[7]Section 1960.35.

[8]Requests for HHEs.

a federal agency health and safety committee, or federal employees not covered by such a committee; or the Secretary of Labor. The legislative authority and limitations are important input to the HHE Program and are also external factors that affect the ability of the program to maximize its impact. Although authority has been broadly interpreted since enactment, NIOSH is legislatively conferred authority for investigation of only a specific subset of occupational health and safety conditions. NIOSH has the ability to compel workplace entry only for toxic substance exposure. When entry is not an issue, the HHE Program evaluates the full spectrum of hazards. The committee infers that the original vision included the notion that chemical exposures that did not violate existing OSHA PELs might be causing health problems that could be identified through medical and exposure evaluation. NIOSH was expected both to intervene in the requesting workplace and to provide data to support new or modified PELs or other health standards.

Over nearly four decades since the act was passed, the scope of hazards that the HHE Program addresses has expanded beyond chemical exposures. The scope of activities to which HHE resources are applied was also extended to provide technical assistance to requesters from other governmental agencies such as OSHA, the U.S. Postal Service, and the U.S. Department of Transportation.

INPUTS

As noted in the HHE Program logic model (Figure 2-1), inputs include planning and the program's use of resources. A challenge facing the HHE Program is balancing the need to be responsive to individual requests against the need to develop strategic approaches to identifying emerging health threats in the workplace. The HHE Program conducted a strategic planning process in the mid-1990s that "focused largely on internal process issues" (NIOSH, 2007b:29). In 2007, the program revised its strategies in response to a 2006 Board of Scientific Counselors (BSC) program evaluation and guidance from the NIOSH Office of the Director (OD).

The mission of the HHE Program is "to protect worker health through problem solving, research, risk communication, and dissemination of findings and recommendations by responding to external requests for hazard evaluations and technical assistance" (NIOSH, 2007b:16). To that end, the program established the following strategic goals:

1. Prevent occupational illnesses through reduced exposure to workplace hazards;
2. Promote occupational safety and health research on emerging issues; and

3. Protect the health and safety of workers during public health emergencies (NIOSH, 2007b:1).

Intermediate goals and performance measures were established in 2007 in response to the BSC 2006 evaluation and are described and assessed in Chapter 3. The HHE Program stated that performance measures will be reviewed annually and revised as appropriate (NIOSH, 2007b).

The strategic goals and activities of the HHE Program are influenced by a variety of factors, including program and agency resources, legislative mandates, HHE requests, and stakeholder needs. The Framework Document (Appendix A) divides inputs into two categories: production and planning. Production inputs (also called program resources in the Framework Document) include program structure and management, funding, staffing, and physical facilities. Planning inputs include input from stakeholders, surveillance and intervention data, and risk assessments.

What follows are brief descriptions of various inputs to the HHE Program. These descriptions are intended to orient the reader to the internal and external factors that shape the program. More detail about each of these inputs is provided as warranted throughout the report.

Production Inputs

Program Organization

The HHE Program summarizes the organization of its activities in the logic model shown in Figure 2-1 (assessed in Chapters 3 and 4 of this report). The program's role in the protection of worker health and safety during public health emergencies is not reflected in the logic model, but is included as an important element in the program's strategic goals. This is also addressed in Chapters 3 and 4.

The HHE Program mission is carried out by employees of the Hazard Evaluations and Technical Assistance Branch (HETAB) in the Division of Surveillance, Hazard Evaluations, and Field Studies (DSHEFS) in Cincinnati, Ohio, and the Field Studies Branch (FSB) in the Division of Respiratory Disease Studies (DRDS) in Morgantown, West Virginia. The administrative lead of the HHE Program resides within HETAB. The HHE Program coordinator also serves as the HETAB branch chief. All HHE requests are logged and tracked within HETAB, which serves as the primary point of contact within the HHE Program for OSHA and for state and local agencies.

There appears to be considerable support from and collaboration with staff throughout NIOSH on specific investigations. The HHE Program described collaboration with NIOSH scientists engaged in many of the priority research areas established by the National Occupational Research Agenda (NORA) including both cross-sector (health outcome) and sector-specific (industry-related) priorities (NIOSH, 2007b). According to the HHE Program, interbranch coordination within the program is facilitated through "joint participation by conference call in regular meetings to discuss and assign incoming HHE requests and by informal exchanges about procedural, policy, and scientific issues" (NIOSH, 2007b).

The HHE Program staff informed the committee that the program regularly partners with the NIOSH Division of Applied Research and Technology and with the Health Effects Laboratory. To fulfill its responsibilities in the area of emergency response, the HHE Program partners with the Emergency Response and Preparedness Branch of the NIOSH OD. In a small number of cases, researchers from other NIOSH programs may be given primary responsibility for carrying out an HHE. When needed technical expertise is unavailable within NIOSH, the HHE Program works with experts in other government agencies or contracts with individuals in the private sector. These collaborations are essential in view of the diversity of potential workplace hazards that require evaluation.

Budget

Table 2-1 shows the HHE Program budget by branch and class over fiscal years 2000 through 2007. The committee did not consider an evaluation of the adequacy or appropriateness of budget allocations as part of its charge, nor was it given the data to conduct such an evaluation. The committee considers funding only in terms of the resources available to the program. Aspects of the program budget are discussed in greater detail as warranted throughout the report text.

HETAB is the administrative home of the HHE Program and bears the costs for program administration. HETAB logs and tracks all HHE requests, maintains a central file of HHE requests and reports, and carries out routine communication functions, such as notifying OSHA and state and local health departments of HHE requests. It also prepares supporting documentation for Office of Management and Budget (OMB) approval pursuant to the Paperwork Reduction Act.

The NIOSH OD allocates funds to the DSHEFS and DRDS, which then allocate funds to their respective branches (NIOSH, 2007b). Approximately $7.7 million were allocated to the HHE Program in 2007 (approximately 3 percent of the total NIOSH budget). Approximately 77 percent of the funding covers personnel costs, while the remainder is intended for discretionary spending. During the past 10 years, most of the program budget has come from the NIOSH base budget, though some funding has been received from earmarks, to offset expenses arising from emergency response activities, or was received from NORA research funding to support core activities.

TABLE 2-1 Budget for the HHE Program by Branch and Class: Fiscal Years 2000-2007

	Class (thousands of dollars)										
Year	Salaries	Benefits	Travel	Mail	Mobile Communications	Printing	Contracts	Supplies	Equipment	Fellows	Total
Hazard Evaluations and Technical Assistance Branch											
2000	3,431	1,047	316	11	18	65	55	87	44	99	5,173
2001	3,529	1,115	314	6	19	26	107	86	130		5,332
2002	3,340	1,203	335	23	34	10	529	145	774	174	6,567
2003	3,866	1,381	376	34	37	6	234	108	124		6,166
2004	3,583	1,275	360	27	14	6	304	114	57		5,740
2005	3,360	1,124	265	20	7		374	129	10		5,289
2006	3,253	1,084	203	18	4	2	199	85	199		5,047
2007	3,502	1,176	198	25	2	1	366	74	105		5,449
Field Studies Branch											
2000	935	232	128	8	0	4	430	34	147	40	1,959
2001	1,324	320	218	1	9	1	359	36	111	61	2,441
2002	1,402	335	256	1	14	0	476	68	49	54	2,655
2003	1,372	350	180	0	12	1	401	31	100	65	2,512
2004	1,382	414	160	5	2	5	446	32	64	0	2,509
2005	1,393	496	158	0	3	3	365	65	5	0	2,489
2006	1,188	395	92	13	3	0	284	59	21	48	2,101
2007	1,195	400	109	0	5	1	210	85	7	21	2,033

SOURCE: NIOSH (2007g).

Personnel

As shown in Figure 2-2, there has been a substantial reduction in the total number of full-time equivalents (FTEs) since a peak in 2001. Some of this decline can be attributed to reorganization by the Centers for Disease Control and Prevention (CDC) involving clerical staff, but staff reductions within HETAB alone approach 20 percent. Periodic staff increases have occurred since 2001, commensurate with specific projects, such as those related to flavorings. The number of FTEs in 2007 was 61 (NIOSH, 2007b).

Staff members are generalists in the areas of occupational medicine, epidemiology, industrial hygiene, ergonomics, engineering, behavioral science, pulmonary and other areas of medicine, toxicology, occupational health nursing, and statistics. Communication with HHE Program staff indicates program staff are distributed by discipline or role within HETAB and FSB in the disciplines needed to conduct HHEs, especially given the collaborative efforts within NIOSH (NIOSH, 2007g).

Table 2-1 indicates a relatively flat budget in real dollars for the HHE Program over the past eight years. Personnel costs have remained nearly constant over this period of time, while contract costs have shown considerable variation. Thus, it appears that the relatively flat budget can be associated with decreases in FTEs for the program over the same time.

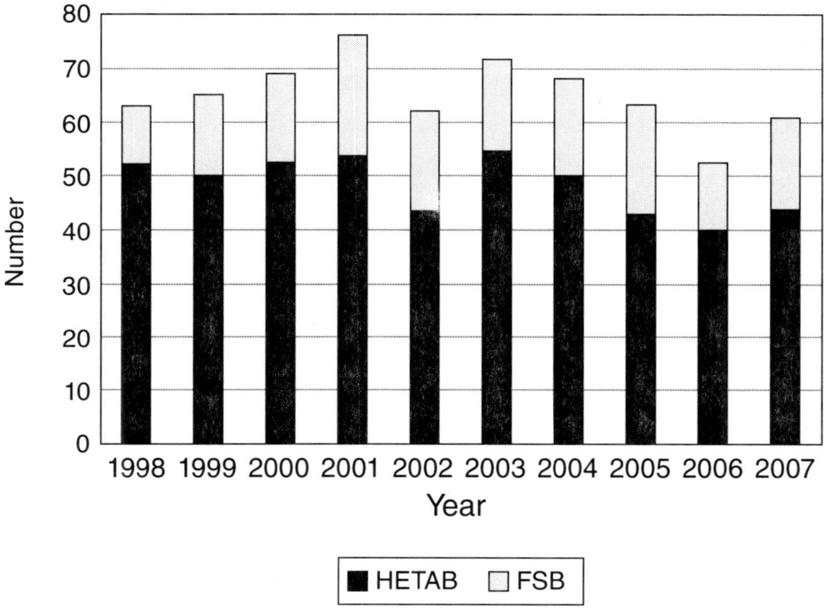

FIGURE 2-2 Total HHE Program FTEs.
SOURCE: NIOSH (2007b).

Facilities

The HHE Program maintains offices and facilities for the storage and maintenance of scientific equipment used in field investigations at both HETAB and FSB facilities. Equipment includes sampling and monitoring equipment for exposure assessment, personal protective equipment for HHE investigators, and logistical support equipment and supply. A trailer designed as a staging facility for field investigations during emergency response and large-scale nonemergency field investigations is maintained at the program's Cincinnati facilities. A medical trailer in Morgantown is equipped to assess pulmonary effects of workplace exposures.

The HHE Program does not maintain its own suite of chemical and biological analytical equipment but relies on resources elsewhere in NIOSH or on NIOSH-funded contract laboratories for sample analyses.

Planning Inputs: Previous Evaluations

Between 1972 and 2006, the HHE Program received planning input in the form of 12 internal and external evaluations (NIOSH, 2007b). Evaluation methods included expert review panels, key informant interviews, and customer satisfaction surveys. The three most recent evaluations were conducted by the Research Triangle Institute (RTI, 1997) and the NIOSH BSC (1997, 2006). While the committee reviewed these evaluations as an aid to understanding the evolution of the HHE Program and its elements, it was not part of the committee's charge to provide an assessment of these evaluations. Nonetheless, pertinent elements of the prior evaluations and the HHE Program responses are discussed in more detail in Chapters 3 and 4.

The 1997 review, developed through a contract with RTI, elicited feedback from a wide range of stakeholders. Problems or classes of criticisms identified include

- lack of timeliness,
- procedural bias (specifically, concerns that employers had greater input into the process than employees),
- overemphasis on routine investigations, and
- lack of practicality in recommendations (NIOSH, 2007b:Appendix 2.3, 14).

RTI recommendations led to the establishment of the HHE followback program, which was fully implemented in October 1999. The goals of the followback program are to "(1) provide feedback to improve the process by which HHEs are conducted and improve the worksite-specific outputs of the HHE Program; and

(2) assess the effectiveness of HHEs in effecting change in the workplace and in improving the health of workers" (NIOSH, 2007b:23). Because of financial constraints that limit followback field investigations, followback evaluations consist of self-reported satisfaction questions to management and employees in the facility served, yielding primarily process rather than outcome information. Questionnaires are distributed for all field investigations and for 50 percent of those receiving technical assistance or consultation. Followback field investigations are conducted for a very limited number of HHEs per year, resources permitting. This is an ongoing process intended to assist the HHE Program with its evaluation of customer service, relevance, and impact. In response to another RTI recommendation, compendia of HHE reports on lead, tuberculosis, noise, and isocyanates were developed (NIOSH, 2001a, 2001b, 1998c, and 2004e, respectively).

The 1997 BSC review was conducted at the request of the NIOSH director (NIOSH, 2007b). Recommendations were made related to

- problems identifying the agenda of the requestor, including labor-management difficulties and work organization issues;
- personnel training needs, especially communication;
- the need to better prioritize HHE requests;
- documentation of HHE impacts, including effectiveness of recommendations;
- the need to identify emerging hazards, less routine work, and more useful summaries of investigations in areas of more extensive experience; and
- maximizing the ability to accomplish the preceding recommendations within the constraints of limited resources.

The HHE Program responded with significant changes to address the second and third of these issues. To improve communication, the program restructured its numbered HHE reports into a uniform format that clearly identifies authorship and responsible personnel. The first page of the reports includes a plain-language discussion of recommendations for employers and employees, followed by a clear summary. Letter reports written by the HHE Program have not been similarly structured, although authorship and telephone contact information facilitate communication.

The second major response was to formally introduce a "triage system" to prioritize incoming requests for HHEs. A mechanism has been developed that allows the HHE Program to respond to routine questions through telephone and fax-back information sharing when program leadership determines that a field investigation would not increase the body of knowledge on a given topic. The triage process itself was described to the committee by HHE Program staff. The committee notes that the criteria for triage, apart from the straightforward assessment of the validity of the request as determined by regulation, do not explicitly outline how HHE personnel determine whether field investigations are necessary or distinguish complex investigations or those yielding research opportunities.

In further response to BSC (NIOSH BSC, 1997) recommendations, the HHE Program developed and twice updated a procedures manual used as a reference for all program staff. Report dissemination, via websites, announcements in NIOSH eNews, announcements to all state epidemiologists and epidemic intelligence officers through Epi-X,[9] NIOSH alerts, and summary documents on particular topics, was enhanced. The ability to triage routine requests by definition frees up scarce resources for enhanced focus on emerging issues; however, metrics to assess the impact of these changes were not formally introduced until after a subsequent BSC program evaluation in 2006.

The 2006 evaluation by the BSC was undertaken at the request of the NIOSH director (NIOSH, 2007b). Recommendations include

- restating HETAB and DRDS missions and increasing the amount of prioritization of HHE requests (triage) to reduce the number of open projects per project officer—the responsibility for responding to routine indoor environmental quality (IEQ) requests was recommended to be removed from the HHE Program;[10]
- improving the efficiency of processes, prioritization of tasks, and overall management of the program, including development of annual goals and performance measures, production of annual reports to demonstrate program outcomes, and utilization of outside consultants;
- identification of the critical mass of staff and specific disciplines necessary to perform essential functions;
- promotion of the HHE Program more widely in an effort to capture more emerging issues and selection of HHEs that will serve program goals; and
- inclusion of estimated health-related savings as a result of HHE recommendations.[11]

Given the short amount of time since these recommendations were made, the HHE Program has not had the opportunity to respond fully to many of them. The program has developed annual goals and performance measures, which are discussed and evaluated in Chapter 3 of this report. Discussions among HHE

[9]Epi-X is a secure electronic communication network maintained by CDC for public health agencies nationwide.

[10]Note: This committee does not fully agree with this recommendation.

[11]Note: This committee questions the feasibility of this recommendation.

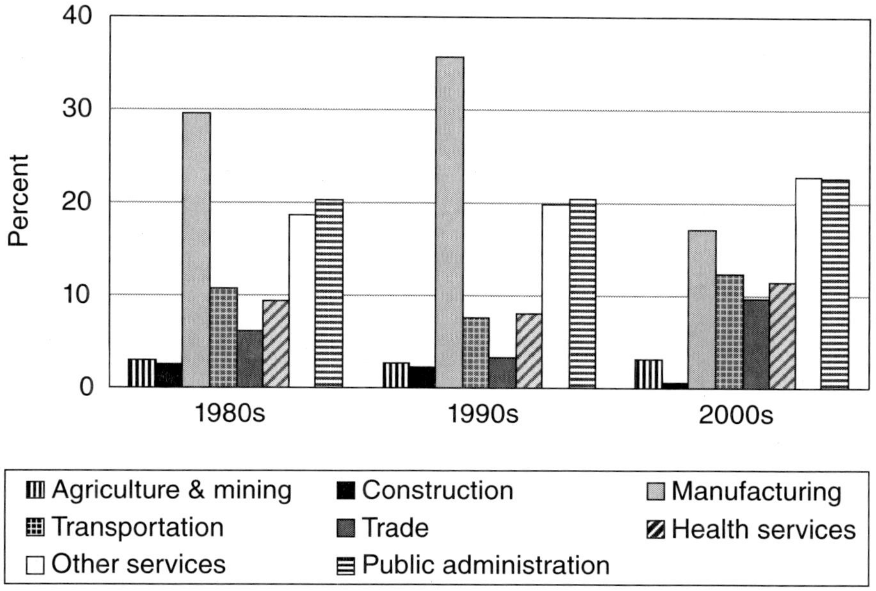

FIGURE 2-3 HHE requests by decade by SIC sector.
SOURCE: NIOSH (2007b).

Program managers have taken place regarding how to proceed with others of these recommendations.

HHE Requests

As already discussed, the HHE Program receives requests for assistance in addressing workplace health issues from stakeholders and applies triage criteria to determine whether a field investigation is warranted. The annual number of HHE requests has remained relatively flat for the past 10 years, with an average of approximately 370 requests each year (NIOSH, 2007b:35). Figure 2-3 shows the distribution of requests made by decade, organized by Standard Industrial Classification (SIC).[12] The distribution of HHE requests by sector has changed over time. For example, from the 1970s through the 1990s, the manufacturing sector accounted for 30 to 40 percent of HHE requests. Today, manufacturing accounts for less

[12]The SIC system (replaced by the North American Industry Classification System [NAICS] in 1997) was developed for use by federal statistical agencies to classify business establishments for statistical analysis purposes (http://www.census.gov/epcd/www/naicsdev.htm, accessed April 16, 2008). HHE Program staff provided data to the committee referring to the SIC system. For the sake of consistency, the committee opted to use SIC rather than NAICS.

than 20 percent (NIOSH, 2007b). Table 2-2 shows the frequency of investigations compared by NORA industry sectors and hazard (NIOSH, 2007g).[13] Comparing the two sectors with the largest number of requests (services and manufacturing), manufacturing-sector requests were more likely to have resulted in field investigations than service-sector requests.

The HHE Program also reported a change in the nature of requests over time. The number of IEQ-related requests, for example, significantly increased following a 1992 CBS evening news segment that encouraged viewers with IEQ problems to contact NIOSH (NIOSH, 2007b). More recently, IEQ requests have typically been made with regard to nonindustrial worksites, such as government buildings, private-sector offices, schools, healthcare facilities, and hotels and resorts (NIOSH, 2007b).

ACTIVITIES

Response to Requests

The HHE Program responds to requests for HHEs or consultation and technical assistance in the form of letter or telephone responses and field investigations. HHE Program managers generally make decisions regarding the appropriate response to incoming HHE requests. As described earlier, the HHE Program has developed a process to prioritize incoming HHE requests and direct resources and responses. The decision processes for triage and response are summarized by HHE Program staff in Figure 2-4 (NIOSH, 2008c). HETAB and FSB managers and supervisors regularly meet to specify under which of four response categories an HHE request may fall (see Box 2-2), following criteria outlined in the HHE Program Procedures Manual (NIOSH, 2006c) as described by program staff and summarized in Box 2-3 (NIOSH, 2007f).[14] Contact with the requestor may be

[13]Table 2-2 was derived by the HHE Program in response to information requests from the committee (NIOSH, 2007h). The data were drawn from the program's management tracking system, developed primarily for record retrieval and not data analyses. Because most HHE requests are related to multiple hazards or hazard classes, the HHE Program developed the following strategy for the tabulation of hazard class of requests:

- IEQ—any request that included an IEQ issue, regardless of other issues also present
- Physical—if not IEQ, and if keywords were present indicating ergonomic, musculoskeletal, radiation, or noise issues
- Biological—if not assigned to IEQ or Physical, and keywords were present relating to tuberculosis, histoplasmosis, biosolids, or brucellosis issues
- All Other—any other request not assigned to the classes above

[14]Box 2-3 represents the HHE categorization strategy per the 1994 procedures manual, which was superseded by a new procedures manual produced in 2006. The 1994 criteria were used over most of the time period being evaluated.

TABLE 2-2 Frequency of HHE Field Investigations by Sector and Hazard: Fiscal Years 1998-2007[a]

	Hazard[c]									
	IEQ		Physical		Biological		All Other		Total	
Sector[b]	N (%)[d]	Total N[e]	N (%)[d]	Total N[e]	N (%)[d]	Total N[e]	N (%)[d]	Total N[e]	N (%)[d]	Total N[e]
Agriculture, Forestry, & Fishing	2 (29)	7	5 (63)	8	1 (50)	2	10 (34)	29	18 (39)	46
Construction	0	14	2 (29)	7	1 (50)	2	30 (47)	63	33 (38)	86
Healthcare & Social Assistance	23 (8)	305	13 (42)	31	4 (36)	11	24 (15)	161	64 (13)	508
Manufacturing	15 (13)	119	36 (47)	76	2 (50)	4	149 (33)	455	202 (31)	654
Mining	2 (66)	3	0	1	1 (50)	2	9 (31)	29	12 (34)	35
Services	108 (9)	1,203	44 (39)	113	12 (48)	25	128 (22)	572	292 (16)	1,913
Transportation	8 (5)	147	12 (25)	48	16 (84)	19	51 (24)	214	87 (20)	428
Trade	1 (1)	129	9 (29)	31	0	1	11 (11)	100	21 (8)	261
Total	159 (8)	1,927	121 (38)	315	37 (56)	66	412 (25)	1,623	729 (19)	3,931

[a]Includes all HHE requests with a completed response as of November 20, 2007.
[b]Sector is based on SIC codes or NAICS codes, depending on year of the request.
[c]See text for an explanation of hazard classes.
[d]Percentage of HHE requests resulting in a field investigation.
[e]Number of HHE requests.
SOURCE: NIOSH (2007g).

for referral to a different agency or to obtain additional information, assist in the formulation of a valid request, or provide readily available information regarding the request. Alternatively, HHE staff may decide that a site visit is warranted, which may result in a letter report with findings and recommendations applicable to the given worksite or a numbered report that may be applicable to other workplaces.

The number of field investigations declined from 126 in 1997 to 58 in 2006 (NIOSH, 2007b). HHE Program staff explained the decrease as due, in part, to the

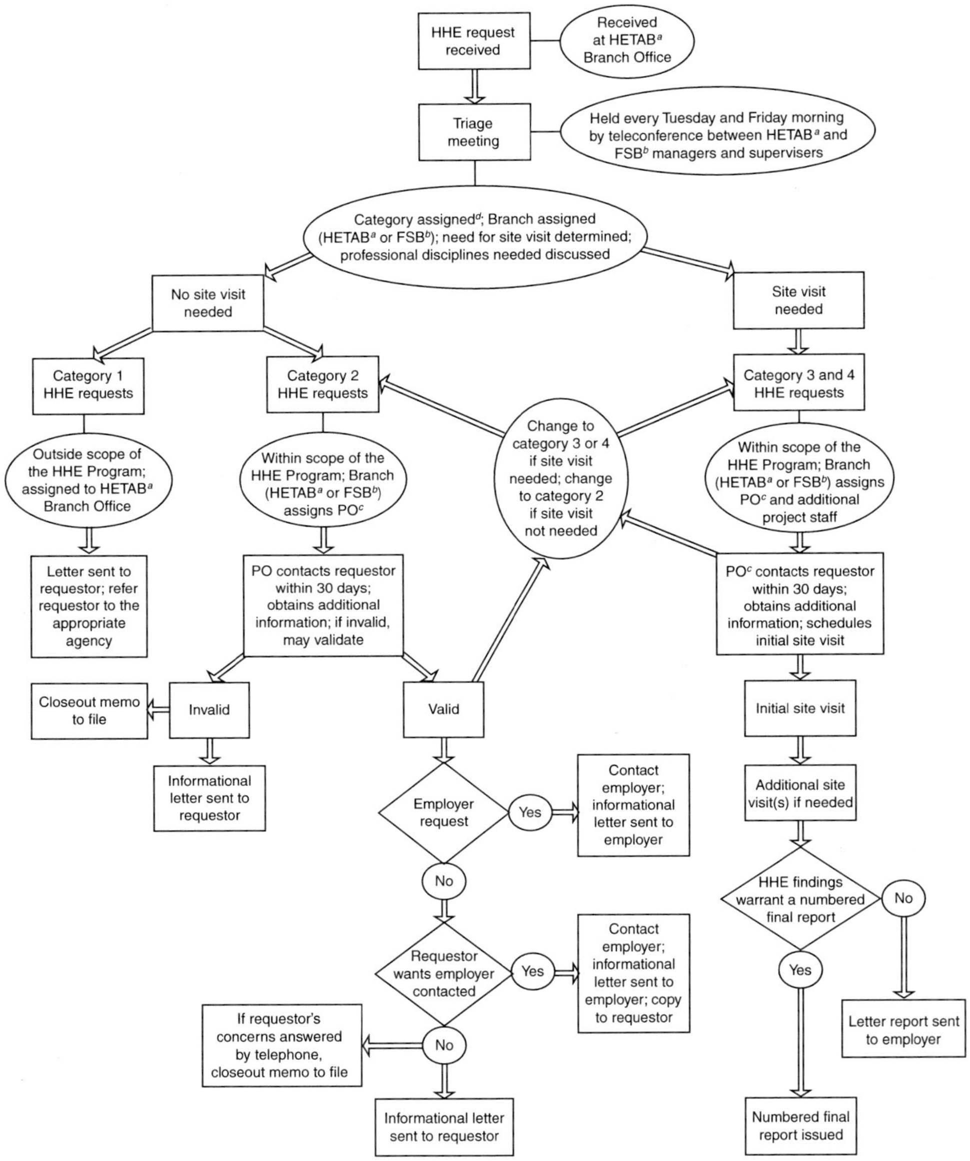

FIGURE 2-4 Flow diagram representing the HHE Program triage process for prioritizing HHE requests and allocating resources for response.
[a] Hazard Evaluations and Technical Assistance Branch (of NIOSH)
[b] Field Studies Branch (of NIOSH)
[c] Project Officer
[d] See Box 2-2 for category descriptions
SOURCE: NIOSH (2008c).

BOX 2-2
Triage Response Categories

In order to prioritize HHE requests, the HHE Program will assign each written request for assistance in one of the following described categories that drive the program response:

Category 1. These requests do not meet the criteria for a valid request [per legislation], concern issues outside the scope of the HHE Program, and are therefore administratively invalid. The HHE Program notifies the requestor and refers the requestor to another agency, such as OSHA or a state or local health department.

Category 2. These are valid or invalid requests for which technical information is supplied to the requestor without conducting a field investigation. Examples include well-recognized problems with readily apparent solutions, problems that have already been adequately evaluated by NIOSH or others. When the request is invalid but the HHE Program believes that a field investigation might be warranted, the requestor is contacted quickly and provided information about what constitutes a valid request.

Category 3. These are valid requests for which a field investigation is necessary to adequately evaluate the occupational safety and health problem described.

Category 4. These are valid requests that present a complex problem or an opportunity for research. These may take longer than Category 3 requests to complete due to required method development or other technical issues.

SOURCE: NIOSH (2007b).

program's ability to respond without field investigations to requests for which there are well-established control solutions. Additional factors attributed to the changing response pattern include the decreased size of HHE Program staff, decreased discretionary funding, increased complexity of field investigations, and increased demands for staff involvement in activities other than traditional HHEs. These activities include emergency response and preparedness, participation on agency and expert committees, document development and review, international technical assistance, and mentoring and training of non-NIOSH occupational health and safety professionals (NIOSH, 2007b).

The HHE Program may conduct a followback survey at a facility where an HHE has been conducted. The surveys are distributed among management and employees and are designed to obtain feedback to improve the HHE process and assess the impact of the HHE in the workplace.

BOX 2-3
HHE Program Triage Criteria

Guidelines for categorizing and prioritizing HHE requests are summarized in a procedures manual (NIOSH, 2006c). Classification is based on information provided by the requestors and is largely dependent on the expertise of HHE Program staff and management. The following excerpted text was provided to the committee in response to questions regarding program practices and represents the categorization strategy as defined by the HHE Program in 1994 (NIOSH, 2007f). Although this strategy was superseded in 2006 by a revised procedures manual, it was applicable over much of the time frame evaluated by this committee.

HHE Categorization Strategy

1. Valid request? If the request is invalid, the requestor will be given the criteria for a valid request and given the opportunity to resubmit the updated request.
2. Does the request relate to a NIOSH special initiative or research project? For example, violence in the workplace, agricultural or construction industries would be placed in Category III or IV.
3. What type of exposures are present in the workplace? Unfortunately, the information included in the original request is frequently incomplete. Therefore, requests from industries with known exposure to extremely hazardous substances should be placed into Category III. In addition, new industries, processes, or exposures not previously evaluated by the HHE Program, NIOSH in general, or other occupational safety and health professionals should be placed into Category III or IV. On the other hand, requests for compliance or routine monitoring would be placed into Category II with referral suggestions (e.g. OSHA). Finally, processes previously evaluated by the HHE Program and found to have minimal opportunity for hazardous exposures should be categorized as either Category II or III.
4. Are adverse health effects being reported? Like the exposure information, information about alleged health effects is frequently misleading. Considerations used to categorize requests based on health effects are the 1) type of condition, 2) severity of the condition, 3) reversibility of the condition, and 4) amenability to control measures. If the condition has been evaluated by a physician, has a diagnosis been made? Severe symptoms causing hospitalization should be categorized into III or IV. Mild symptoms should be categorized into II or III. Previously unrecognized health effects from common exposures, or occupational groups not known to be at risk for a particular health effect should also categorized as III or IV. Requests intended only to provide medical or industrial hygiene evidence for a pending legal action should be placed in Category II.

5. Is the workforce unique? For example, is the workforce composed of only women, only minorities, migrant labor, etc. If yes, Category III should be considered.

Criteria for Prioritizing HHE Requests in Category III

1. Does the request relate to a NIOSH special initiative or research project? For example, violence in the workplace, agricultural or construction industries.
2. Can the evaluation be done by someone else? Examples: Companies can hire private-sector consultants. Government agencies can use Federal Occupational Health (FOH); some agencies have their own health and safety staff. OSHA requesters can consult with their national Technical Assistance Directorate in Washington, DC. Some state health departments have the resources to investigate. In some cases, an OSHA technical consultation (for management) or inspection (employee) may be more appropriate than an HHE.
3. Who is the requestor? Employee, small business, and union requests will be given a higher priority given their resource and access limitations.
4. What types of exposures are present in the workplace? Requests from industries with known exposure to extremely hazardous substances should be given a higher priority than other Category III requests. In addition, new industries, processes, or exposures not previously evaluated by the HHE Program, NIOSH in general, or other occupational safety and health professionals should be ranked higher. Industries or processes previously evaluated by the HHE Program should be ranked lower.
5. Are adverse health effects being reported? Considerations used to prioritize requests based on health effects are similar to those used to categorize requests. Therefore, the 1) type of condition, 2) severity of the condition, 3) reversibility of the condition, and 4) amenability of control measures are important to consider. Has the condition caused the affected employees to seek medical care? Severe symptoms causing hospitalization should be given high priority. Mild symptoms should be categorized into II or III. Previously unrecognized health effects from common exposures, or occupational groups not known to be at risk for a particular health effect should be given higher priority. Requests intended only to provide medical or industrial hygiene evidence for a pending legal determination should have a low priority.
6. Geographic. In times of critical shortage of travel funds, is the request close to Cincinnati, Morgantown, or one of our regional offices, therefore incurring minimal travel expenses?

Emergency Response

The HHE Program's responsibilities in response to disasters and emergency preparedness have changed dramatically since the terrorist attacks on September 11, 2001 (9/11), and the program's role is still evolving. Because of HHE Program staff's experience in assembling rapid field response, they provided leadership for NIOSH activities in the aftermath of the 9/11 attacks on the World Trade Center and for response to the subsequent anthrax contamination through the U.S. postal system. Fifteen HHE Program staff, along with other NIOSH staff, responded to the immediate protection needs of rescue and recovery workers in New York City. Such workers included firefighters, law enforcement personnel, emergency medical services, construction and demolition trades, health and safety personnel, volunteer workers, and local, state, and federal agency workers. The HHE Program provided technical assistance to local, state, and federal agencies related to personal protective equipment; assisted with air sampling to characterize the complex environment; and conducted several specific HHEs (NIOSH, 2007b).

HHE Program staff described the training and resource development involved in preparing for their role in emergency response settings. The HHE Program provides staff training, personal protective equipment and fit testing, and appropriate medical evaluations, using an all-hazards model. The model includes biological, chemical, explosive, and radiological events, as well as natural disasters, including hurricanes, earthquakes, and wildland fires. The HHE Program participates in tabletop exercises that include some staff training with partners in Top Officials (TOPOFF), a terrorism preparedness exercise involving officials from all levels of government, representatives from the international community, and the private sector.[15]

Before 9/11, the HHE Program responded to emergencies and managed the activities of deployed program staff. Subsequently, NIOSH and HHE Program roles in response have been prescribed by the National Response Framework (U.S. Department of Homeland Security, 2008).[16] The HHE Program does not serve as the lead agency, but it has a defined responsibility. The program is one of the biggest contributors of technical leadership, but the organizational structure falls under the NIOSH OD. HHE Program staff members have often been asked to serve as team leaders because of their extensive field experience in difficult situations, for

[15]Department of Emergency Management and Homeland Security (http://www.ct.gov/demhs/cwp/view.asp?a=2017&q=290966&pp=12&n=1, accessed July 8, 2008).

[16]The National Response Plan was replaced by the National Response Framework, effective March 22, 2008. The National Response Framework presents the guiding principles that enable all response partners to prepare for and provide a unified national response to disasters and emergencies. It establishes a comprehensive, national, all-hazards approach to domestic incident response (http://www.dhs.gov/xprepresp/committees/editorial_0566.shtm, accessed July 8, 2008).

example, in the response to Hurricane Katrina in 2005. Many of the people staffing the CDC operations center were HHE Program staff.

After 9/11, the HHE Program received substantial funding for a truck and trailer for field deployment, and it is responsible for their maintenance and use. The program also received funding to construct a garage to house the vehicle and for staging purposes. The equipment is housed at the program's Cincinnati facility.

Pre- and post-deployment medical follow-up have been provided for program staff deployed in emergency response activities. Telephone guidance and other resources have been made available to assist staff with mental health issues when considered necessary (NIOSH, 2008a).

OUTPUTS

HHE Program outputs include official numbered reports and less formal letter reports in response to requests for HHEs and technical assistance. The program also produces articles for peer-reviewed publications, website content, and presentations at professional, trade, and agency conferences and meetings. The primary means by which the program communicates its findings is through numbered HHE reports. The current format of numbered reports includes a page of HHE "highlights" written in lay terms, which includes information about the HHE request, program activities and findings, and bulleted lists of what managers and employees can do to minimize investigated hazards. A more technical summary follows the highlights section. The body of the report then provides background, investigation methodologies and criteria, results, discussion, conclusions, recommendations, and a list of references. Test results are also made available.

Technical assistance and letter reports can be very similar in scope to numbered HHE reports or may be much more focused and answer specific questions or provide referrals as necessary.

Between 1996 and 2006, the HHE Program produced 495 numbered HHE reports, 503 letter reports for other field investigations, and 1,999 technical assistance or consultation letter reports (NIOSH, 2007b). The committee reviewed several numbered and letter reports provided by the HHE Program, as well as several HHE reports obtained elsewhere. These and other HHE Program outputs are discussed and evaluated at greater length in Chapters 3 and 4.

3

Relevance of the Health Hazard Evaluation Program

EVALUATION OF RELEVANCE

As discussed in Chapter 1, a relevant program can be described as an integrated program involving interrelated surveillance, research, and transfer activities. The relevance of its activities can be determined by the severity or frequency of hazards addressed and the number of people at risk, the extent to which the program addresses gender-related issues and those of underserved populations (see Box 1-1 for definition), and the extent to which the health and safety needs of small businesses are addressed. Program structure and content must also increase program relevance.

The Health Hazard Evaluation (HHE) Program primarily serves a legislated, public health practice mission, and its relevance must be evaluated in that context. HHEs address issues that, by definition, are in the early stages of investigation and intervention. One important function of an HHE is to provide preliminary data that may lead to a more extensive research program. The mandate of the program, however, requires it to respond to requests, whether or not more in-depth research may follow. Within this constraint, the committee believes the program can elicit and prioritize requests to fit into an interrelated surveillance, research, and transfer matrix involving the rest of the National Institute for Occupational Safety and Health (NIOSH). One of the challenges faced by the HHE Program is weighing the need to be responsive to individual requests with the need to develop strategic approaches to identifying emerging health threats in the workplace. A careful balance serves both functions. Through response to requests the program may identify emerging issues, and identifying emerging issues helps in priority-setting among investigations.

To assess program relevance, the committee begins this chapter with a review of major inputs to the program, including the strategic planning and stakeholder processes, and considers whether resources have been appropriately allocated and used to meet strategic needs. The chapter continues with an evaluation of the HHE Program's activities and outputs, a discussion of how the program handles emerging issues, and then a discussion of the relevance of the program's emergency response activities. The committee's findings related to each of these topics are highlighted in bold. Recommendations related to the improvement of program relevance are found within the body of the text throughout the chapter and are organized and summarized in Chapter 5. The final section of this chapter provides the committee's rating for relevance.

RELEVANCE OF INPUTS (PLANNING AND RESOURCES)

Strategic Planning

Finding 1: The HHE Program's strategic plan is highly relevant to the program mission.

The HHE Program's strategic and intermediate goals are summarized in Table 3-1. These were developed by the HHE Program in response to a recommendation by the NIOSH Board of Scientific Counselors (BSC) to "improve the efficiency of processes, prioritization of tasks and overall management of the program [including to ... d]evelop annual goals and measurable objectives regarding the work products, priorities and work processes of the HHE Program that are practical, cost effective, and consistent with resources" (NIOSH BSC, 2006). The committee finds that the HHE Program has made a serious effort to respond to this recommendation. The goals are well targeted and relevant, and can be measured by the number and types of HHE requests received, the number of reports and field investigations conducted related to each of these goals, and their effects as determined through followback surveys (described in Chapter 2). Appropriate performance measures for each intermediate goal have been chosen, given the limited resources of the program. The committee believes, however, that with additional resources, performance measures could be made more specific and ambitious. There is no indication that surveillance or workers' compensation data were used to formulate the HHE Program's strategic plan. Given more resources, onsite follow-up investigations to assess the impact of interventions would help the program assess and revise its goals and performance measures as necessary. The committee comments on individual performance goals and measures in Table 3-1.

TABLE 3-1 HHE Program Strategic and Intermediate Goals, Performance Measures, and Committee Comments

Strategic goal 1: Prevent occupational illnesses through reduced exposure to workplace hazards.

Intermediate Goals	Performance Measures	Committee Comments
1.1. Increase the number of HHE requests concerning important occupational health problems	**1.1.1. Annual goal: Conduct a targeted marketing campaign** 1.1.1.1. FY 2007: Award one contract for a customer survey 1.1.1.2. FY 2007: Stimulate two valid HHE requests through outreach to physicians in the Cincinnati area **1.1.2. Annual goal: Enhance awareness of the HHE Program** 1.1.2.1. FY 2007: Increase the percentage of national union websites that include a link to the HHE Program website (obtain baseline data) 1.1.2.2. FY 2007: Increase the percentage of state Occupational Safety and Health Administration (OSHA) websites that include a link to the HHE Program website (obtain baseline data) 1.1.2.3. FY 2007: Establish links to the HHE Program website on the website of three government agencies or trade associations providing services to small business 1.1.2.4. FY 2007: In the 16 states in which an HHE was announced in Epi-X in FY 2004-2006, assess awareness of the HHE Program among state epidemiologists	The committee endorses this intermediate goal, which is the core of the HHE Program mission. In addition to enhancing awareness, appropriate performance measures include establishing partnerships to develop valid requests from small businesses and underserved worker populations. Partnerships with nonprofit organizations, community development organizations, and worker centers could be explored, and specific metrics concerning requests could be generated and tracked. The use of international, national, state, and local surveillance data would facilitate identifying and focusing resources in areas of emerging need and underserved populations.
1.2. Facilitate implementation of recommendations at facilities investigated through the HHE Program	**1.2.1. Annual goal: Provide a service that is valued by HHE participants** 1.2.1.1. FY 2007: 80% of respondents to the HHE followback survey report that the NIOSH evaluation was "good" or "excellent" one year after the report was issued (baseline: 82% for FY 2001-2005)[a] **1.2.2. Annual goal: Provide a service that results in improved workplace conditions** 1.2.2.1. FY 2007: 80% of respondents to the HHE followback survey report that workplace conditions have improved since the NIOSH evaluation (baseline: 81% for FY 2001-2005)[a]	While the committee agrees there is a need for continued commitment to process evaluation, return site visits that offer the ability to provide objective evaluation of impact are critically important, though adequate resources may need to be obtained. The committee notes a decrease in field investigations, which may at least partially explain the decrease in response time for

1.2.3. Annual goal: Produce timely reports for HHE requests addressed with a field investigation 1.2.3.1. FY 2007: Complete 50% of reports within 6 months of the last site visit (baseline: 36% for investigations with a last visit in FY 2005)[a]	investigations that are being conducted. The timeliness of the responses to HHE requests is critical to the success of the program. With the diminishing number of HHE requests, the program should be able to set more timely response goals.
1.2.4. Annual goal: Produce useful reports for HHE requests addressed with a field investigation 1.2.4.1. FY 2007: 95% of respondents to the HHE followback survey report that the Recommendations section of the report was "very or somewhat useful" to them (baseline: 94% for FY 2001-2005)[a] 1.2.4.2. FY 2007: 95% of employees responding to the HHE followback survey report that the Highlights section of the report was "very or somewhat useful" to them (baseline: 96% for FY 2001-2005)[a]	Follow-up with technical assistance after report issuance should be carefully evaluated for knowledge transfer and exchange to other relevant facilities.
1.2.5. Annual goal: Provide continuing technical assistance after the issuance of the final report 1.2.5.1. FY 2007: Make onsite, post-investigation presentations for two HHEs 1.2.5.2. FY 2007: Develop a partnership with one facility to assist in implementation and evaluation of recommendations	
1.2.6. Annual goal: Assess implementation of HHE recommendations through followback surveys 1.2.6.1. FY 2007: Carry out a mailed survey regarding recommendations 1 year after issuing a final report for 100% of onsite HHEs 1.2.6.2. FY 2007: Successfully obtain information from 70% of participants in the mailed survey 1.2.6.3. FY 2007: Complete at least four onsite followback surveys	

continued

TABLE 3-1 Continued

Intermediate Goals	Performance Measures	Committee Comments
Strategic goal 1: Prevent occupational illnesses through reduced exposure to workplace hazards.		
1.3. Provide appropriate technical assistance for HHE requests addressed without a field visit	**1.3.1. Annual goal: Provide timely informational letters** 1.3.1.1. FY 2007: Complete 90% of letters within 6 months of the request (baseline: 87% for FY 2005 requests)[a] **1.3.2. Annual goal: Provide helpful informational letters** 1.3.2.1. FY 2007: 65% of HHE requestors completing a followback survey report that the letter was helpful (baseline: 60% for FY 2001-2005)[a] 1.3.2.2. FY 2007: 65% of HHE requestors completing a followback survey report that they shared information in the letter with others at the worksite (baseline: 59% for FY 2001-2005)[a] **1.3.3. Annual goal: Develop standard informational materials for common concerns** 1.3.3.1. FY 2007: Two draft indoor environmental quality (IEQ) NIOSH-numbered documents will be submitted for external peer review	Performance measure 1.3.3 would greatly free up resources while informing a wide range of workplaces about ways to reduce poor IEQ. It should be possible to produce similar educational materials for other areas of high-frequency requests in which the HHE Program has considerable experience.
1.4. Employees and employers at facilities not investigated through the HHE Program are aware of hazards identified and controls recommended by HHE Program investigators	**1.4.1. Annual goal: Enhance the dissemination of HHE reports** 1.4.1.1. FY 2007: Announce all HHE reports in eNews and Epi-X within 2 months of distributing to the investigated facility 1.4.1.2. FY 2007: Post HHE reports on the NIOSH website within 2 months of distributing to the investigated facility **1.4.2. Annual goal: Develop communication products based on HHEs (e.g., NIOSH-numbered documents, trade publications, and presentations)** 1.4.2.1. FY 2007: Prepare and disseminate 1-2 new written products 1.4.2.2. FY 2007: Present HHE findings at 2-3 trade meetings 1.4.2.3. FY 2007: Prepare and disseminate an annual report of highlights from the HHE Program	This intermediate goal should further increase program relevance. Additional avenues for disseminating information may include video-sharing websites or public service announcements. County agriculture extension services may be used to disseminate agriculture-related findings. Similarly, some states have engineering extension services to work with small manufacturing facilities and economic development councils.

1.5. Professional practices, guidelines, policies, standards, and regulations are influenced by information generated from the HHE Program	**1.5.1. Annual goal: Transfer information generated from the HHE Program to other occupational safety and health professionals** 1.5.1.1. FY 2007: Submit 4 to 6 peer-reviewed manuscripts based on HHEs 1.5.1.2. FY 2007: Make 20 to 25 presentations based on HHE Program work at scientific or agency meetings 1.5.1.3. FY 2007: Participate as requested in agency workgroups	The committee applauds the HHE Program's recognition of the critical role it can play in knowledge transfer, and encourages an increased focus on knowledge exchange to increase relevance in identifying emerging issues.

Strategic goal 2: Promote occupational safety and health research on emerging issues.

Intermediate Goals	Performance Measures	Committee Comments
2.1. Customers will have information about emerging issues	**2.1.1. Annual goal: Use the internal HHE review process and HHE project officer surveys to identify emerging issues** 2.1.1.1. FY 2007: Prepare and disseminate an annual list of new findings from HHEs	The approach described could be augmented by developing a tickler file of unusual requests for which no clear findings have been made. These could be reviewed periodically in a process that includes intramural and extramural scientists to identify new patterns.
	2.1.2. Annual goal: Transfer information regarding emerging issues to appropriate internal and external partners 2.1.2.1. FY 2007: Complete and accurate information regarding new HHE requests and closed projects is entered in the HHE database 2.1.2.2. FY 2007: Prepare and disseminate annual HHE summary data to National Occupational Research Agenda (NORA) research councils for the manufacturing, construction, service, and healthcare and social assistance sectors 2.1.2.3. FY 2007: Provide timely summary information generated from the HHE Program to all NORA sectors	Interaction with NORA sector councils will increase recognition of the HHE Program within these industries as well as the HHE Program's understanding of potential hazards within each industry. Sectors chosen by the HHE Program are certainly relevant, but the program could also consider being represented on the NORA transportation council. This industry has high injury and illness rates, particularly within trucking, and few HHEs have been conducted ($n = 3$, with the last one in 1989).

continued

TABLE 3-1 Continued

Intermediate Goals	Performance Measures	Committee Comments
Strategic goal 3: Protect the health and safety of workers during public health emergencies.		
3.1. Partners and customers have the information they need regarding high-priority occupational health issues likely to arise during public health emergencies	**3.1.1. Annual goal: Develop and disseminate information anticipating likely chemical, biological, or radiological threats** 3.1.1.1. FY 2007: Complete 100% of tasks related to pandemic flu planning according to schedule	The HHE Program has been effective in this arena and should communicate the importance of this role to ensure resources are available to continue these activities.
3.2. HHE Program personnel respond appropriately to requests for assistance	**3.2.1. Annual goal: Maintain adequate numbers of trained and certified staff** 3.2.1.1. FY 2007: A minimum of 25 personnel are enrolled in a respiratory protection program, maintain Hazardous Waste Operations and Emergency Response Standard (HAZWOPER) certification, complete incident command system training, and are aware of information resources regarding specific potential hazards 3.2.1.2. FY 2007: Assurance is received from the Office of Health and Safety that HHE emergency responders have appropriate medical oversight	Emphasis should be placed on recognizing policy implications of HHEs in this arena. Explicit training should be given on how to recognize and appropriately seek guidance on handling HHE requests with broader policy implications.

3.3. Ensure that the HHE Program is ready to respond to requests for assistance	**3.3.1. Annual goal: Emergency response-related equipment is available for field deployment within 24 hours of request to respond** 3.3.1.1. FY 2007: The electronic inventory of industrial hygiene equipment is complete and current 3.3.1.2. FY 2007: Calibrations are up-to-date for 100% of industrial hygiene equipment likely to be used in an emergency deployment, and records of current calibration dates are maintained 3.3.1.3. FY 2007: The Hazard Evaluations and Technical Assistance Branch (HETAB) emergency response vehicle is appropriately equipped and maintained, records are up-to-date, and a minimum of two drivers have commercial drivers licenses 3.3.1.4. FY 2007: At least three HHE Program staff members are currently certified to ship dangerous goods **3.3.2. Annual goal: Management systems are in place to facilitate efficient response to requests for assistance** 3.3.2.1. FY 2007: Database of responders is current and complete with regards to home, work, and cellular telephone numbers for responder and emergency contact, respirator fit test results, and security clearance status 3.3.2.2. FY 2007: Develop standard operating procedures for communications and preparation for responding	This is an important goal. The committee recognizes the need for sufficient resources.

[a]Based on the three most recent years for which data are available.
SOURCE: NIOSH (2007b:Appendix 2.7, HHE Program Strategic Plan).

Recommendation: Conduct regular assessments of performance measures to determine whether available resources allow more ambitious goals.

Use of Resources

The Framework Document (Appendix A) directs the committee to consider how planning, production, and other input data support and promote program goals and activities, and to determine whether input is obtained from stakeholders, including from underserved populations and small companies. Planning inputs include surveillance data; advice from advisory committees, National Occupational Research Agenda (NORA) research partners, and other stakeholders; reports from the Fatality Assessment Control and Evaluation (FACE) program; and extramural health outcome, exposure assessment, or similar data from federal, state, and other program partners. Production inputs, including budget, staff, facilities, and management, play major roles in the program.

Finding 2: For the most part, the HHE Program has judiciously used its resources to meet its mission in the face of the changing economy, the changing nature of HHE requests, and increased responsibilities related to emergency response.

HHE Program activities have been modified over the past 10 years because of events such as the September 11, 2001 (9/11) attacks; specific HHE topics such as diacetyl; the influence of the NORA process; and response to program evaluations. The 1997 BSC evaluation (NIOSH BSC, 1997) prompted the formalization of the HHE Program triage process,[1] described in Chapter 2, and the procedures manual, which have resulted in improved program efficiency and other improvements.

Health Hazard Evaluation Requests and Resource Allocations

Figure 3-1 shows the change in the number of HHE requests and program responses between 1997 and 2006. The number of requests during this period peaked in 2001 at about 580. At the end of the time frame, HHE requests declined to approximately 390, which is only approximately 15 percent more than in 1997. By the end of the period, more responses to HHE requests took the form of written

[1]Triage is the system for allocating scarce resources to cases or problems based on a priority scheme, especially in emergency medicine and public health. The term comes from battlefield medicine where the wounded are separated into three categories—those who are likely to die even with care, those who are likely to recover even without care, and those for whom care would be effective.

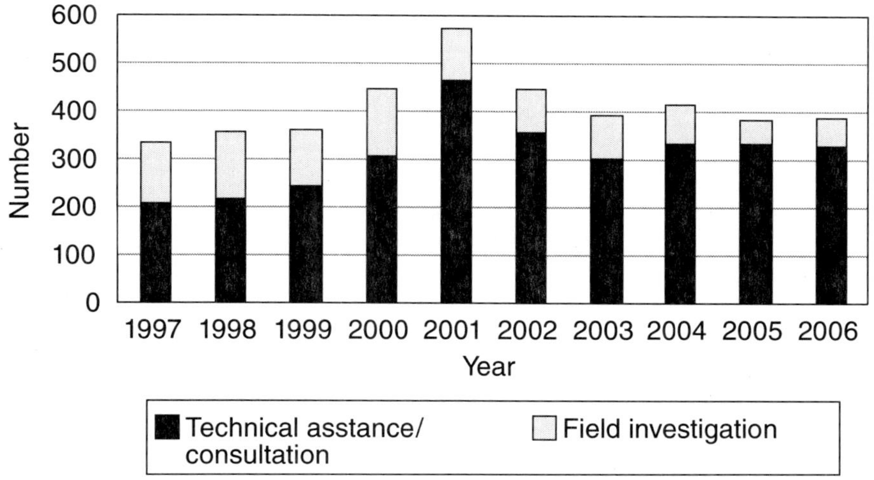

FIGURE 3-1 Distribution of HHE requests by response category.
NOTE: Category 1 and 2 requests are combined under technical assistance/consultation. Category 3 and 4 requests are combined under field investigation (see Box 2-2 for a description of the HHE request categories and responses).
SOURCE: NIOSH (2007b).

information and telephone communication rather than field investigations. These forms of follow-up were specifically developed in response to external reviews and indicate HHE Program success in redirecting resources to provide appropriate information in response to routine questions. However, this increased proportion of routine requests may also suggest that the HHE Program now has the opportunity to conduct more outreach and encourage more requests (for example, those that meet strategic needs or represent new or emerging hazards) that would require site visits (if sufficient funding is available to conduct them).

The committee asked the HHE Program to provide information regarding all HHE requests, including types of requestor, hazard and sector involved, and the types of responses. Tables 3-2 and 3-3 were derived by the HHE Program in response (NIOSH, 2007g). The data were drawn from the program's management tracking system, which was developed primarily for record retrieval and not data analyses. Because most HHE requests are related to multiple hazards or hazard classes, the HHE Program developed the following strategy for tabulation of hazard class of requests:

- Indoor environmental quality (IEQ)—any request that included an IEQ issue, regardless of other issues also present

- Physical—if not IEQ, and if keywords were present indicating ergonomic, musculoskeletal, radiation, or noise issues
- Biological—if not assigned to IEQ or Physical, and keywords were present relating to tuberculosis, histoplasmosis, biosolids, or brucellosis issues
- All Other—any other request not assigned to the classes above

In its response to the committee, the HHE Program recognized the limitation of this methodology.

Table 3-2 lists the type of HHE response by source of request, hazard, and percentages of resulting field investigations by requestor type between 1998 and 2007. During this time, approximately half the requests were related to IEQ issues and were largely addressed without field investigations. Among the five sectors with more than 100 requests, the correlation between the percentage of field investigation responses and the percentage of IEQ requests is –0.82. There were very few requests related to biological issues; thus it was appropriate that the greatest proportion of requests resulting in field investigations occurred in this area. Many requests in the physical hazards category are likely related to musculoskeletal hazards, which is appropriate given the magnitude of the problem in most industries. The committee wanted to explore trends in requests, both with and without consideration of IEQ requests, to determine if those requests potentially skew the data. The HHE Program databases could not provide the data in a manner that would allow more robust analysis.

The effect of the changing economy can be seen in Figure 2-3, which indicates a greater than 50 percent decline in the number of requests from the manufacturing sector since the 1990s. One might expect this change to allow the program greater flexibility to respond more efficiently to the increasing number of HHE requests in health services, trade, and transportation and to conduct unexpected emergency response activities. Telephone, fax, and written responses to HHE requests have increased in the past 10 years (NIOSH, 2007b). The percentage of field investigations generated by IEQ and service-sector requests is very low compared with those from manufacturing (see Table 2-2).

To facilitate efficiency when responding to requests, HHE Program staff review available reports from other parties who have investigated the issue or site and provide opinions regarding the methods, data interpretation, and appropriateness of conclusions and recommendations. HHE Program staff may then gather additional information from employer and employee representatives during the triage process to determine the request category and HHE Program response. HHE Program staff consult with other NIOSH scientists, review the scientific literature, and prepare written responses summarizing their activities. These steps improve efficiency and provide consistency in HHE outputs.

TABLE 3-2 Frequency of HHE Field Investigations by Source of Request and Hazard Class: Fiscal Years 1998-2007[a]

	Source of Request (Requestor)													
	Employee		Government[c]		Management		Union		Joint		Other		Total	
Hazard[b]	*N* (%)[d]	Total *N*[e]	*N* (%)[d]	Total *N*[e]	*N* (%)[d]	Total *N*[e]	*N* (%)[d]	Total *N*[e]	*N* (%)[d]	Total *N*[e]	*N* (%)[d]	Total *N*[e]	*N* (%)[d]	Total *N*[e]
Biological	1 (8)	12	26 (87)	30	8 (53)	15	2 (40)	5	0	0	0	5	37 (55)	67
IEQ	62 (5)	1,332	17 (33)	51	49 (22)	223	24 (13)	189	4 (31)	13	3 (3)	118	159 (8)	1,926
Physical	35 (22)	158	14 (67)	21	45 (73)	62	18 (38)	47	7 (100)	7	2 (10)	20	121 (38)	315
All Other	102 (13)	817	89 (54)	164	116 (45)	260	82 (39)	214	16 (73)	22	6 (4)	147	412 (25)	1,624
Total	200 (9)	2,319	146 (55)	266	218 (39)	560	127 (28)	455	27 (64)	42	11 (4)	290	729 (19)	3,932

[a]Includes all HHE requests with a completed response as of November 28, 2007.
[b]See text for an explanation of hazard classes.
[c]This group may include some HHE requests more appropriately classified as management due to a change in coding over time.
[d]Percentage of HHE requests resulting in a field investigation.
[e]Number of HHE requests.
SOURCE: NIOSH (2007g).

The committee notes that the percentage of field investigations tends to vary by requestor type. For example, Table 3-2 indicates there were large numbers of employee requests ($n = 2{,}319$), of which 9 percent ($n = 200$) resulted in field investigations, whereas 39 percent of the far fewer management requests ($n = 560$) resulted in field investigations. More than 1,300 of the employee requests were for IEQ issues. As described below, NIOSH appears to have used appropriate discretion in providing extensive information on control measures to the vast majority of IEQ requestors in the form of technical assistance and consultations.

Table 3-3 addresses the type of HHE response by source and sector. There were very few requests from agriculture and mining. The proportion of requests resulting in field investigations was less for healthcare and social assistance (13 percent) and services (15 percent) than for most other sectors. While this is likely related to IEQ requests identified in Table 3-2, these sectors may be growing and also reflect a higher proportion of underserved populations, again potentially indicating the need for the program to assist stakeholders in developing meaningful and valid requests. Similarly, industries generating few requests might benefit from HHE Program exploration into the reasons so few requests are made followed by appropriately directed outreach activities.

As noted in Table 3-1, the committee endorses the HHE Program's intermediate goal of increasing the number of HHE requests concerning important occupational health problems (intermediate goal 1.1). The committee agrees that an increase in valid requests will increase the scope and penetration of the HHE Program into more industries and allow better transfer of information to more workers, including those from underserved populations and those employed by small businesses. During discussions with the committee, HHE Program staff appeared aware that with an increase in the number of requests comes the potential for increased constraints on resources.

Program Resources and Emergency Response

In the last decade, the HHE Program has received a number of requests that involved assessing risks, measuring exposures, and providing guidance to workers and the general public during disasters (NIOSH, 2007b). These assignments often involved the onsite presence of HHE staff working in concert with partners from the Federal Emergency Management Agency (FEMA), the Department of Health and Human Services Department of Emergency Response, the Center for Disease Control and Prevention (CDC) National Center for Environmental Health, the CDC Director's Emergency Operations Center (DEOC), and several global agencies and manufacturers. Response to domestic disasters, such as Hurricanes Katrina, Floyd, and Isabel, and to massive flooding required mobilization of up to 18 HHE

TABLE 3-3 Frequency of HHE Field Investigations by Source of Request and Sector: Fiscal Years 1998-2007[a]

	Source of Request (Requestor)													
	Employee		Government[c]		Management		Union		Joint		Other		Total	
Sector[b]	N (%)[d]	Total N[e]	N (%)[c]	Total N[d]	N (%)[e]	Total N[d]	N (%)[c]	Total N[d]	N (%)[c]	Total N[d]	N (%)[c]	Total N[e]	N (%)[d]	Total N
Agriculture, Forestry, & Fishing	4 (25)	16	6 (55)	11	8 (4)	13	0	1	0	0	0	5	18 (39)	46
Construction	4 (14)	28	4 (67)	6	19 (66)	29	6 (50)	12	0	0	0	11	33 (38)	86
Healthcare & Social Assistance	21 (7)	319	11 (42)	26	23 (28)	83	7 (18)	39	2 (40)	5	0	36	65 (13)	508
Manufacturing	69 (21)	355	17 (47)	36	39 (51)	76	60 (57)	105	16 (84)	19	1 (2)	62	202 (31)	654
Mining	0	3	4 (31)	13	2 (67)	3	6 (38)	16	0	0	0	0	12 (34)	35
Services	73 (6)	1,146	73 (53)	139	98 (34)	289	33 (18)	185	5 (39)	13	10 (7)	140	283 (15)	1,913
Transportation	18 (7)	243	26 (90)	29	24 (53)	45	15 (17)	90	4 (80)	5	0	16	87 (20)	428
Trade	11 (5)	207	5 (83)	6	5 (23)	22	0	7	0	0	0	19	21 (8)	261
Total	200 (9)	2,319	146 (55)	266	218 (39)	560	127 (28)	455	27 (64)	42	11 (4)	289	729 (19)	3,931

[a]Includes all HHE requests with a completed response as of November 20, 2007.
[b]Sector is based on Standard Industrial Classification codes or North American Industry Classification System codes, depending on the year of the request.
[c]Number of HHE requests.
[d]Percentage of HHE requests resulting in a field investigation.
[e]This group may include some HHE requests more appropriately assigned to the management group due to a change in coding over time.
SOURCE: NIOSH (2007g).

Program staff responders, smaller numbers of public health professionals for onsite investigations, or long-distance support from HHE Program staff through conference calls (NIOSH, 2008b). HHE Program staff also responded to events such as 9/11 and the subsequent anthrax investigations.

Emergency response activities create competing demands on limited resources within the HHE Program. Routine program operations continued during these unplanned activities, but significant coordination and additional outside funding were required to maintain continuity. Requests for assistance during disasters have so significantly impacted the workload of HHE Program staff that the BSC was prompted to recommend major process changes to improve efficiency (NIOSH BSC, 1997). The triage process and a procedures manual developed in response to the evaluation have created flexibility and efficiency to effectively respond to public emergencies while maintaining normal program operations. More discussion on HHE Program emergency response activities is found in a later section of this chapter (Finding 9).

Recommendation: Continue to provide guidance and recommendations during public health emergencies. The HHE Program would benefit from the development of a mechanism to reduce the impact of emergency response activities on routine program functions.

Expertise

With the exception of pulmonologists on staff in the Field Studies Branch (FSB), the HHE Program, out of necessity, is staffed primarily by generalists (NIOSH, 2007b) and must rely on resources beyond the HHE Program for specialized expertise. HHE Program staff informed the committee how necessary expertise is identified during the triage process and indicated that many areas of expertise are readily available through existing, relatively informal arrangements within and external to NIOSH. It was not made clear how certain areas of expertise, such as dermatology or clinical toxicology, are accessed, or how HHE staff acquire information in such areas as health services or intervention effectiveness research. Nor was it made clear whether there was explicit understanding of the economic environment of a worksite—important for assessing the economic feasibility of recommended control solutions.[2] The HHE Program's ability to utilize expertise

[2]The BSC recommended that the HHE Program include estimates of costs and savings associated with implementing HHE recommendations (NIOSH BSC, 2006). The committee agrees that feasibility is an important consideration when making recommendations. The expertise needed to provide cost estimates, however, does not reside within the program, and providing such estimates could delay dissemination of results and potentially detract from the HHE Program mission.

from other parts of NIOSH could be enhanced by a formal mechanism to contract externally for needed clinical, toxicologic, or other specific expertise, especially as resources become limited and experienced individuals retire. Formalized relationships for cooperation may be helpful, but should be negotiated carefully to ensure continued flexibility when people, expertise, or analytical services beyond the terms negotiated are needed.

A program that addresses the wide array of issues that the HHE Program does, including during emergency situations, would not be successful without residing in and having access to the expertise in NIOSH. A review of the authorship of HHE reports and subsequent peer-reviewed journal articles indicates that the HHE Program uses expertise from elsewhere in NIOSH for a wide variety of occupational health issues, and that successful collaborations have resulted in peer-reviewed publications. However, as in many other agencies, programs, and institutes that employ public health professionals, many NIOSH senior professionals are approaching retirement. The HHE Program and NIOSH could prepare for this shift through succession planning, contemplated reassignments, and identification of key resources needed in the near future (see the committee's discussion of the training of new professionals in Chapter 4).

Finding 3: The HHE Program has not made sufficient use of available surveillance data to assist in targeting field investigations to recognize previously unknown hazards or to identify new or increased hazards caused by changes in the workforce and work environment.

The committee was informed that the HHE Program uses traditional methods, such as results from investigations, literature reviews, case files, or other internal documents, and contacts within the scientific community to identify emerging, previously unknown, or increased hazards in the workplace.

There is no indication that surveillance or workers' compensation data are used to formulate the HHE Program's strategic plan, or that intervention effectiveness data are examined prior to making recommendations. The program's strategic plan may have been influenced by surveillance data, given the extent to which NORA and the overarching NIOSH strategic goals may be influenced by surveillance data. There is also little evidence that the HHE Program conducts environmental scans and routinely monitors various state and federal databases (for example, OSHA's Integrated Management Information System database [IMIS] or Bureau of Labor Statistics [BLS] databases), reports, or websites on which surveillance information may be available to identify potential emerging hazards (see Box 3-1). An environmental scan is more than a review of injury and illness statistics. It takes into account societal and economic changes and pressures, demographic changes, and

BOX 3-1
Surveillance and the HHE Program

A strong national occupational health surveillance system could be of benefit to the HHE Program, which itself is not a surveillance program. Greater HHE Program relevance and national impact may be achieved if HHE Program findings were incorporated into a national surveillance system in which active searching and reporting of trends and clusters are possible. The HHE Program could influence NIOSH management regarding the importance of a national surveillance system, perhaps implemented in the Surveillance Branch of the Division of Respiratory Disease Studies (DRDS). The NIOSH Sentinel Event Notification System for Occupational Risk (SENSOR) program might be used to feed data to the system. A good example of an existing system is the reporting of asthma by providers to the Massachusetts Occupational Health Surveillance Program (Massachusetts Office of Health and Human Services, 2008). Access to comprehensive surveillance data could allow the HHE Program to more effectively prioritize HHE requests, make more informed triage decisions, and identify emerging workplace health hazards.

the ways industry uses new technology and the workforce, as well as how government action and international trade may affect hazardous exposure in an industry. Perhaps HHE Program physicians routinely participate in occupational health professional meetings or electronic dialogues from which they receive informal information regarding trends observed in occupational medicine clinics and in field investigations. They may then suggest HHE requests as a means to a solution. The extent to which such informal channels are used is not known.

The committee was not provided evidence that the HHE Program conducts detailed explorations of workers' compensation or other data to gain insight into the magnitude and severity of an issue or to reveal circumstances that could contribute to specific hazards. For example, a significant increase in proportionate mortality from respiratory disease among cooks was identified in the Washington State Department of Health Occupational Mortality Database (Washington State Department of Health, 2008). Such systems may be useful to identify industries or workplaces in need of investigation or to identify those who need information regarding the control of hazards. Additionally, outreach to workers' compensation carriers and loss control consultants may generate new requests for HHEs from clients for which hazards are identified.

HHE Program staff members are engaged in professional meetings and national conferences but more could be done to reach a greater diversity of employers and industries. The HHE Program could use such opportunities to play a more active role in identifying potential hazards in emerging industries or processes. For example, a formal program of seminars, discussion groups, or other mechanistic science routes might enable the HHE Program to link hazardous substances to various industrial uses and to identify groups of exposed workers with adverse health effects, more than would routine monitoring of national and international databases.[3] Furthermore, the HHE Program could keep track of chemicals used in the workplace and the potential for hazardous exposure through other sources. Spencer and Schaumburg (2000), for example, list a large number of neurotoxic materials. The HHE Program could then actively inform companies, worker advocates, and healthcare professionals who may be aware of exposures or apparent health effects in specific populations about the program and request process to encourage evaluation of these exposure situations.

Recommendation: The HHE Program should make systematic use of professional meetings, scientific conferences, scientific literature, and surveillance data, including those generated by NIOSH, to assist in prioritizing field investigations and recognizing emerging issues.

Finding 4: The extent and effectiveness of relationships between the HHE Program and federal and state agencies are variable.

The committee observed both positive and inadequate interactions of the HHE Program with state and federal agencies.

Through formal and informal reporting relationships with state departments of health, labor, unemployment, environment, and workers' compensation programs, the HHE Program may help identify worker clusters that could benefit from the conduct of HHEs. Trends analyses in unemployment and industry-sector shifts, including identification of pockets of underserved contingent workers, might enhance the HHE Program's ability to reach high-risk populations. Maximizing HHE Program potential in this area, however, depends on the ability of the program to maintain appropriate staffing levels. These relationships may also allow the program to better communicate its mission and activities and to provide better and timelier feedback.

Three examples demonstrate ways to maximize federal and state resources and to enhance the HHE Program's ability to perform its mission. During the committee's third meeting (see Appendix C for agenda), a stakeholder from the California Department of Public Health spoke with the committee regarding the long-standing and strong relationship between California's Occupational Health

[3]Such databases include those of the Environmental Protection Agency, OSHA, the National Center for Environmental Health, the International Agency for Research on Cancer, and the Health and Safety Executive in Great Britain.

Branch (OHB) and the HHE Program (Materna, 2008). The relationship has resulted in capacity building at OHB in a number of issues, including most recently, those associated with diacetyl (see Box 3-2 for a discussion of diacetyl). During the same meeting, management of a worksite at which an HHE was conducted (NIOSH, 2007i) discussed how useful HHE recommendations had been in the eventual elimination of diacetyl in their processes. In this case, three organizations worked cooperatively to maximize resources and the result was a positive impact.

BOX 3-2
Bronchiolitis Obliterans (Popcorn Lung) and Diacetyl

In 1985, the HHE Program conducted an investigation in an Indiana food processing plant where two employees were diagnosed with bronchiolitis obliterans, a severe and sometimes fatal lung disease (NIOSH, 1986). Investigators failed to determine the causative agent and recommended general industrial hygiene controls. In 1999 and 2000, several current and former workers from a Missouri microwave popcorn plant were diagnosed with bronchiolitis obliterans, and some of these workers were awaiting lung transplants. Bronchiolitis obliterans is rare in the general population, so the workers' physician notified the state health department, which inspected the facility in March of 2000 for risks to public—but not worker—health (Michaels et al., 2008).

In May 2000, the health department contacted the Occupational Safety and Health Administration (OSHA) in Kansas City, described an alleged cluster of bronchiolitis obliterans, and requested an inspection for compliance with regulations. The request was made of OSHA because it was felt that OSHA could "address this situation, and if there is an obvious hazard to workers, address it quickly" (Roberts, 2000). The health department did not have the statutory authority to inspect the facility without a clear determination of a health hazard. An OSHA inspector visited the plant and focused the inspection on well-recognized respiratory hazards: dust and oil mist. The inspector declined to sample for dust because the company's insurance carrier had done so earlier. Oil mist samples were collected, but OSHA lab methodologies were incompatible for use with vegetable oils. Although OSHA was aware of the bronchiolitis obliterans cases, the OSHA area office determined the company to be in compliance and closed the file (OSHA, 2000).

In August 2000, the Missouri health department contacted NIOSH. Over the next several months, the HHE Program conducted comprehensive industrial hygiene sampling and a health assessment of current employees. The investigation found that rates of adverse respiratory symptoms were significantly higher than in the general population and that the likely cause was the artificial butter flavoring diacetyl (NIOSH, 2006b). In December 2000, NIOSH issued interim recommendations to help prevent exposure to harmful flavorings (NIOSH, 2000). Over the next three years the HHE Program revisited the plant every four to six months to conduct follow-up air sampling and medical testing and to evaluate the effectiveness of the controls it recommended (NIOSH, 2006b). Following its initial investigation at the Missouri plant, the HHE Program also conducted evaluations in other locations (Kanwal et al., 2006), issued an alert that provides guidance to the industry and its workers (NIOSH, 2003e), and generally worked to better understand

Though there are many successes, as described above, stakeholders from both local and state public health departments told the committee of the need for more information about the HHE Program. The commissioner of health of a major East Coast city health department told the committee he knew nothing about the HHE Program or of the multiple HHEs conducted in his city over the years (Sharfstein, 2007). A Massachusetts Occupational Health Surveillance Program industrial hygienist indicated her personal hesitancy in filing HHE requests for

and communicate the risk to workers who produce or apply flavorings, their employers, and the occupational health community.

Diacetyl has generated widespread media attention, perhaps because it is a food additive with exposures possible not only among workers who manufacture flavorings, but also among consumers. The Food and Drug Administration (FDA) was alerted in 2007 that a consumer was diagnosed with severe lung disease, whose clinical findings and exposure levels were noted to be similar to workers diagnosed with bronchiolitis obliterans (Rose, 2007). An investigative reporter looked beyond flavor manufacturing for possible harmful worker exposure downstream among professional cooks (see, e.g., Schneider, 2008). HHEs are pending on diacetyl exposure among professional cooks, but none have been conducted to date. The King County, Washington, Health Department has requested fact sheets for cooks that their restaurant inspectors can distribute during restaurant inspections (Washington State Department of Labor & Industries, 2008a).

In a report from the U.S. House of Representatives Committee on Appropriations dated July 2007, concern was expressed about the potential hazards to workers and consumers exposed to diacetyl. NIOSH data were cited as compelling evidence of a real threat (U.S. Congress, House of Representatives, Committee on Appropriations, 2007:98). According to that report, the FDA should conduct further studies to examine the safety of diacetyl. In response to public concern and pressure from organized labor, OSHA announced in September 2007 that it would initiate rule-making (Steenhuysen, 2007). In that same month, the U.S. House of Representatives passed a bill entitled the Popcorn Workers Lung Disease Prevention Act, directing OSHA to issue a standard regulating worker exposure to diacetyl (U.S. Congress, House of Representatives, House Education and Labor, 2007), which is currently on the Senate calendar. At the state level, legislation was introduced in California to ban the use of diacetyl in the workplace, but the bill has been put into an inactive file of the California Senate (California Assembly, 2007). As of January 2008, four major popcorn manufacturers had voluntarily decided to remove diacetyl from their products (Associated Press, 2007; Schneider, 2007).

The example of diacetyl and bronchiolitis obliterans illustrates the HHE Program's unique ability to conduct epidemiological investigations and identify previously unknown hazards. OSHA and other agencies do not have this same authority or capacity. The program's findings and recommendations have had an immediate effect on the worksites investigated, have instigated research in other parts of NIOSH, have influenced voluntarily changes within industry to remove the hazards, and have had at least some influence on policy setting and rule-making.

other asthma-related issues "because, in some cases HHE requests have resulted in long, drawn out processes, in which the referring state agency may not be apprised of the process, local expertise is not used, the report is not issued for a long time, and the end results may not address the most compelling questions in a timely way" (Pechter, 2008). These comments suggest the need for both outreach and ongoing communication.

Better CDC acknowledgement of the depth and breadth of the occupational health expertise available to state health departments through the HHE Program may result in greater cooperation between the HHE Program and states in identifying emerging occupational health issues. For example, the recent identification of a new form of progressive inflammatory neuropathy among workers processing pig brains (MMWR, 2008) might have benefitted from earlier attention to the work-relatedness of the initial complaints. A number of prior CDC investigations, including the initial evaluation of a case of pulmonary anthrax in Florida (MMWR, 2001a), suggest a pattern of delay in including appropriate occupational health expertise in investigations. The HHE Program could more broadly influence workplace health assessments conducted by other CDC institutes by working more closely with CDC overall.[4] Additionally, public recognition of the HHE Program may be diminished as the 1-800-66-NIOSH telephone number formerly used for contacting NIOSH about occupational safety and health issues is subsumed under the general CDC call line. NIOSH could maintain visibility with other agencies and the public if it were able to reinstate that telephone number.

Recommendation: Enhance HHE Program outreach to OSHA national and regional offices and to state health and labor departments to better communicate the function and activities of the HHE Program, increase cooperation with these agencies, and provide more complete and timely feedback.

RELEVANCE OF ACTIVITIES

Finding 5: The HHE Program has responded well to HHE requests as mandated, although mechanisms for eliciting a broader array of HHE requests are needed.

There is strong evidence the HHE Program has conducted relevant field investigations and has performed well in identifying emerging hazards, serious

[4]HHE Program staff indicated that currently available communication pathways within the CDC (such as Epi-X) may lack timeliness or completeness (NIOSH, 2007c, 2008b).

occupational health problems, and hazards affecting large numbers of workers. Examples of such cases include (1) bronchiolitis obliterans (popcorn lung) associated with diacetyl in the manufacture of flavorings, (2) interstitial pneumonitis and flock exposures, (3) work-related musculoskeletal disorders associated with cumulative trauma, and (4) biological and chemical hazards in the meat and poultry industries. These examples are described more fully in Boxes 3-2, 3-3, 3-4, and 3-5, respectively. All four examples included multiple field investigations and successful transfer of knowledge to the occupational health community by the HHE Program.

BOX 3-3
Interstitial Pneumonitis (Flock Worker's Lung) and Flock

Flock refers to fine, small-diameter synthetic fibers such as those applied to adhesive-coated fabrics to produce a pleasing surface similar to velvet. Breathing flock can cause a serious lung disease called interstitial pneumonitis, which causes inflammation and scarring in the lungs. Flock-related illness was first seen in 1991 at a nylon flock processing plant in Ontario, although the causative agent was not identified. Investigators initially speculated that the condition was caused by mold-contaminated adhesive, which was replaced. However, in 1995 the plant reported two additional cases. In 1995 and 1996, two workers from a Rhode Island plant owned by the same company were diagnosed with interstitial lung disease by the same physician (Lougheed et al., 1995; Eschenbacher et al., 1999). At the physician's urging, the company asked NIOSH to conduct an HHE.

The investigation included industrial hygiene surveys of a variety of potential exposures, respiratory tests, and a medical questionnaire. NIOSH identified flock dust as the causative agent and recommended "decisive, proactive action to install effective engineering controls, to enforce good work practices, to assure appropriate use of proper respiratory protection, to establish a medical screening/surveillance program, and to implement effective administrative controls" (NIOSH, 1996a:33).

Flock has since been implicated in other outbreaks of occupational lung disease and has been the subject of a workshop and recommendations by NIOSH. HHEs have been performed in several other flock processing plants. However, flock never generated the press attention that diacetyl has, perhaps because it is not a food additive and not a hazard to consumers. Flock has never been considered for rule-making by OSHA; the only mention of flock on OSHA's website refers to the explosion hazard related to flock dust (OSHA, 1998). One recent follow-up report indicates that while some industrial hygiene controls have been implemented, flock exposures remain high (NIOSH, 2006a).

BOX 3-4
Work-Related Musculoskeletal Disorders

Work-related musculoskeletal disorders (WMSDs) are disorders of the muscles, tendons, joints, and associated neurovascular structures that occur over time rather than in response to an acute traumatic event. Examples include rotator cuff syndrome at the shoulder, lateral epicondylitis at the elbow (tennis elbow), carpal tunnel syndrome at the wrist, sciatica, and back pain. While these disorders can be unrelated to work activities, their work-relatedness is associated with repetitive work, awkward postures, and overexertion in lifting, pushing, pulling, or carrying, as well as hand-arm vibration. The likelihood of WMSDs increases with the intensity, duration, or frequency of exposure to these risk factors or when multiple risk factors occur simultaneously. These disorders represent 30 to 40 percent of all reported occupational injuries and illnesses, lost work days, and workers' compensation costs. In an example from Washington State, a 26-year-old female nursing assistant employed in an adult family home for seven years developed back problems lifting and transferring residents. She missed 180 days of work and required $5,200 of medical treatment (Silverstein and Adams, 2007).

The HHE Program has played a pivotal role in the area of WMSDs, demonstrating very high relevance for their HHE findings related to hazard identification and abatement. Eighty-two WMSD-related HHEs were conducted between 1980 and 2007 (19 of them between 2000 and 2007). Early HHEs in food processing and other manufacturing brought public, research, and regulatory attention to the area. Experiences gained in the HHE Program resulted in a series of peer-reviewed and practice-oriented publications, including a comprehensive review of the literature on WMSDs (NIOSH, 1997b). The comprehensive review was referenced in the European Union (Buckle and Devereau, 1999), was used extensively in the development of regulations at the national level, and was used to justify the making of ergonomic rules in at least one state[a] (Washington State Department of Labor & Industries, 2000). Many of these HHEs were also used in various rule-making efforts for both evidence of hazards and viable solutions (Washington State Department of Labor & Industries, 2008b), demonstrating a high degree of regulatory relevance. A later, congressionally requested review by the National Academies substantiated the earlier HHE Program work (NRC and IOM, 2001). Other documents based on HHE Program outputs include several manuals to assist employers and employees in the development of effective ergonomics programs, including those intended for underserved populations such as farm workers (NIOSH, 1997a, 2001e, 2007j). These documents have been widely circulated within the health and safety practitioner community, as well as to small and large employer and worker organizations at professional conferences. A number of these ergonomics publications have been adapted and distributed by the HHE Program.

---

[a]Washington State adopted an ergonomics rule in May 2000 (WAC 296-62-05101) (http://www.lni.wa.gov/WISHA/Rules/GeneralOccupationalHealth/PDFs/ErgoRulewithAppendices.pdf, accessed August 9, 2008).

BOX 3-5
Biological and Chemical Hazards in the Meat and Poultry Industries

The HHE Program has identified and addressed several emerging hazards and continues to play a major role in investigating risks in the expanding meat and poultry processing industry, an industry increasingly comprised of a non-union, minority, immigrant, and contingent workforce characterized by low wages and high risk (GAO, 2005b). The HHE Program, working with the State of North Carolina and the U.S. Department of Agriculture (USDA), identified the occupational transmission of brucellosis to workers in swine processing facilities (NIOSH, 2007b). Brucellosis is an infectious bacterial disease that can cause "a range of symptoms that are similar to the flu and may include fever, sweats, headaches, back pains, and physical weakness. Severe infections of the central nervous systems or lining of the heart may occur. Brucellosis can also cause long-lasting or chronic symptoms that include recurrent fevers, joint pain, and fatigue" (NIOSH, 2007b:57). An HHE investigation (NIOSH, 1994) provided the information needed for the USDA to develop and implement a formal rule change that provides compensation for swine herds destroyed because of infection, effectively eliminating the hazard by removing disease-carrying animals from the food chain. This outcome reflects an extraordinary level of cooperation among both federal and state agencies.

In another case, the HHE Program conducted initial and full follow-up investigations of a management-initiated request to evaluate respiratory complaints in a poultry processing area in which hyperchlorinated water was used in the evisceration process (NIOSH, 2003a, 2006g). The report related to an initial HHE investigation demonstrating excess levels of chlorine and trichloramines generated by this process and the subsequent reduction of exposure following elimination of hyperchlorinated water. The final report clearly documents the impact of an aggressive workplace management approach that went beyond the initial HHE Program recommendations and effectively reduced the hazard and improved outcomes in that plant. During the initial investigation, NIOSH engineers encouraged plant management and engineers to take the issue seriously. However, there is no indication that these recommendations have been applied more broadly in the poultry processing industry.

Reported more recently in the news is an outbreak of progressive inflammatory neuropathy among swine slaughterhouse workers (CNN.com/health, 2008; MMWR, 2008). The CDC was called to investigate this as a potential contagious disease. The inclusion of the HHE Program in the investigation has facilitated the broader evaluation of the condition as an occupational illness and further demonstrates the relevance of the HHE Program.

"Casting a Larger Net"

Based on information submitted to the committee, it is not possible to determine whether the needs of underserved populations and small businesses are being met or whether the lack of anonymity or the fear of legal or political reprisals inhibits the request for HHEs. Targeted outreach to state health, labor, and environmental departments; small business and their employees; labor unions; health centers providing care to underserved populations; community and ethnic organizations; and nongovernmental worker centers serving immigrant and contingent workers would increase the likelihood of meeting the needs of these populations. It is especially important to produce educational materials in multiple languages appropriate to these diverse populations. HHE Program outreach would benefit from provisions for assisting requestors, particularly those who are not unionized or who belong to underserved populations, in formulating valid requests.

Communication with non-unionized and underserved populations could be enhanced through collaboration with emerging worker centers that address the needs of new immigrant workers; public service announcements geared toward truck drivers and agricultural and construction workers on radio stations and website links (for example, trucker websites); or targeted outreach to health centers providing care to underserved populations.

The BSC recommendation to "cast a larger net" to enable the program to "select only the evaluations that truly serve program goals" (NIOSH BSC, 2006:5) deserves attention here: while many attempts have been described to increase the overall number of HHE requests, there has been no systematic effort to ascertain either the relative success of these efforts or the reasons for any failures. Specific approaches to identify and encourage new sources of appropriate and valid requests do not appear to have been evaluated. It would be useful if outreach programs could be evaluated for effectiveness.

Ongoing systematic scanning of surveillance data could assist the HHE Program in identifying emerging hazards in underserved populations. While there are no specific plans to identify and control hazards that specifically affect underserved populations, it is possible to focus resources in industries and workplaces that employ a significant proportion of underserved workers.

Recommendation: Establish formal relationships with organizations representing underserved populations, small businesses, and their employees.

Recommendation: Use innovative techniques to reach small businesses and underserved populations, creating a broad array of mechanisms for communicating with diverse constituencies and attending to issues of literacy, language, and national-origin barriers. The effectiveness of applied outreach should be evaluated in a formal manner.

The Health Hazard Evaluation Program Triage Process

Finding 6: The HHE triage process is generally efficient but requires more structure.

Formalization of the Triage and Response Processes

Program evaluations by the Research Triangle Institute (RTI) and the BSC emphasized the need for the HHE Program to conserve scarce resources when addressing its core mission (RTI, 1997; NIOSH BSC, 2006). In response to BSC recommendations, the HHE Program introduced the triage system described in Chapter 2 (NIOSH BSC, 1997). The HHE Program developed structured triage mechanisms that allow the delivery of standard information packages to requestors when full investigations were not warranted (for example, requests triaged as Category 1 or 2; see Box 2-2). Many IEQ-related requests may fall under this category.

The criteria applied during triage, apart from the straightforward assessment of request validity as determined by regulation, do not explicitly outline when field investigations are necessary, nor do they assist in distinguishing those requests that require complex investigations or that may yield future research opportunities. Classification appears to be handled informally. Initial determinations are not always final; project officers can change the category of response to a request after additional information is obtained.

A clear understanding of an HHE request and the implications of a potential response should be explicit in any triage decision. For example, a letter response to a requestor reviewed by the committee documented an attempt by HHE Program staff to provide thoughtful review and useful reference information, but included conclusions based on limited information when resources may have been available elsewhere to address the concerns in greater depth. The report involved a request from management of a multinational petroleum company to explore a possible cluster of esophageal cancer (NIOSH, 2005e). The HHE Program was asked "to evaluate the strengths, weaknesses, and validity of [a prior industry-conducted study at the worksite] and to give suggestions or recommendations for further investigation." A thorough in-house response to this request would have consumed excessive HHE Program resources. The HHE letter report provided a general explanation about the limitations of cancer cluster investigations and recommended no further investigation of the suspected cancer cluster.

Conveying this type of information effectively via telephone and written contact poses a difficult risk communication challenge, but this communication means may be useful when scarce HHE resources are unlikely to provide useful new information for workers. It could be perceived, however, that the HHE Program was used to validate the company's research without sufficient critical perspective being brought to bear. If neutral third-party expertise is required, large entities with adequate resources could be encouraged to develop tripartite oversight (by NIOSH, labor, and industry) of external experts and formally sponsor a thorough review. If the information is potentially of major interest, this could be triaged to other areas of NIOSH.

It is unclear how many letter reports generated by the HHE Program would raise similar issues. The decision against using scarce resources to conduct full site investigations (for Category 3 or 4 requests) should not be used to provide approval or premature conclusions. A description of the needed resources—including, when appropriate, independent extramural research expertise—and alternative approaches to addressing the issue would also be useful. The HHE Program would benefit from a formal mechanism to respond by recommending tripartite studies, funded by the requesting entity when appropriate. General guidelines should exist for referral to higher-level review, particularly for requests involving large companies with multiple resources or requests that would result in the provision of recommendations for which HHE staff may not be sufficiently expert or have the appropriate certifications. When sufficient information about the hazard or issue is readily available, requestors might be referred to the Occupational Safety and Health Administration (OSHA) Small Business Consultation Program or appropriate OSHA directorate (for example, the Directorate of Cooperative and State Programs or the Directorate of Science Technology and Medicine).

Identification of Needed Expertise

A review of several HHE letter reports by the committee indicates that the triage process may not be effective in determining whether staff expertise is suitable for the range of potential program responses to a request. The triage program would ideally involve epidemiologists, toxicologists, engineers, and other relevant specialists, as well as HHE Program staff, to ensure that qualified personnel are handling responses. For example, the committee reviewed a close-out memo from an HHE Program file regarding a state health department request for assistance related to possible worksite metals contamination (NIOSH, 2007a). The memo documented a medical review by HHE Program staff and the provision of diagnostic information by telephone, although the recipient of the information was the public health official who requested the assistance, not a clinician responsible for the individual with the health concerns. HHE Program personnel, of necessity, are generalists. Slightly more than half of HHE personnel have basic certification, such as certified industrial hygienist or medical board certification in a primary specialty, but none have additional toxicological certification (NIOSH, 2007b). Further, the provision of clinical advice requires a therapeutic relationship with a patient. The memo does not state whether the HHE Program provided the individual's treating physician with contact information for an appropriately qualified medical toxicologist for a clinical consultation.[5]

Inclusion of specialists in the triage process would also help identify requests where exposure responses are potentially unusual or not readily explicable on the basis of current knowledge, potentially leading to a higher categorization of the HHE request.

Transparency of the Process

The committee reviewed input from numerous program stakeholders. In general, HHE Program guidance and recommendations were considered well reasoned and sound, but some comments from stakeholders raised issues related to the triage process. Multiple stakeholders did not understand why their HHE requests failed to result in investigations. The committee does not have enough information to determine whether this concern is widespread, but it did determine that greater transparency in the triage process would, at the very least, help the program determine whether the confusion is avoidable. Responding to HHE requests with a systematic approach of defining the request, identifying the implications of potential responses, and identifying justifiable resources and necessary expertise based on the triage criteria would provide the means for the program to better communicate the outcomes of the triage process to requestors. A more transparent triage process could potentially help requestors understand why their requests may be considered invalid and help them form valid requests in the future.

Recommendation: Implement, as part of the triage process, a formal technical assistance mechanism to help requestors to formulate valid HHE requests. In cases where an HHE is not appropriate or where resource limitations prohibit an investigation, technical assistance should include referral to more appropriate NIOSH divisions or government agencies.

Recommendation: Develop an explicit, written process for classifying and prioritizing HHE requests. Priority should be based on the gravity

[5]The American Board of Medical Specialties defines medical toxicology as specialization in the "prevention, evaluation, treatment and monitoring of injury and illness from exposures to drugs and chemicals, as well as biological and radiological agents" (http://www.abms.org/Who_We_Help/Consumers/About_Physician_Specialties/preventive.aspx, accessed September 18, 2008). Three years of specialized training is required to receive a certificate in the subspecialty of preventive medicine.

of the potential harm, the number of employees potentially at risk at similar workplaces or using similar work processes, the urgency of the problem, the potential to assess health outcomes, and the possibility of identifying emerging issues. Potential impact on standards and policy should also enhance the priority of an HHE request in the triage process. Relationship of the HHE to current research may be considered but should not be the only or primary factor. The process should provide guidance on weighting these varying factors.

Recommendation: Better formalize the triage process, including the identification of needed expertise, and improve the transparency of the process to HHE requestors, while maintaining flexibility and speed.

RELEVANCE OF OUTPUTS

Finding 7: HHE reports are generally well written, present relevant information supported by appropriate documentation, and reflect a high level of expertise. However, the committee did not find evidence that a well-defined quality assurance process is in place to ensure consistently high-quality outputs and recommendations.

Review of Numbered Reports

The HHE Program provided the committee with several examples of numbered HHE reports as well as less formal letter responses to HHE requests on a variety of occupational health-related issues. The committee also reviewed several other numbered HHE reports obtained directly from the HHE Program website. The majority of listed authors of numbered HHE reports reviewed held advanced degrees and certifications, which enhances credibility and professionalism. The report formats were uniform and straightforward, with specific recommendations provided in clear language at the outset, followed by a plain-language summary, then the body of the report. All HHE field investigation reports were dated, and nearly all include both the date of the initial request and the source of the request (for example, management, employees, union).

Report timeliness, a major concern expressed in all previous HHE Program evaluations, continues to vary, with the most rapidly delivered reports delivered four to six months from the request date and the majority appearing within a year of the field investigation. One outlier, the final report for an incident that recognized the association between flavorings and lung disease (NIOSH, 2006b), appeared 5½ years after the initial request and 2½ years after the final follow-up visit.

Reasons for the delay were not apparent to the committee. Timeliness, however, was an issue only for the final numbered report. This particular investigation was highly relevant and productive. The hazard was identified and reported to workers and management in a series of timely letter reports. The investigation resulted in new research activities, peer-reviewed publications, and policy efforts. The final report included all results in tabular form and an exhaustive set of appendixes that captured relevant correspondence and other materials, comprising nearly 200 pages of clear information. Many other reports delivered more quickly also contained clear information, tables of exposure and outcome data, and relevant references.

Public accessibility to more complete data is extremely important for program relevance and impact. It would be useful to determine whether the traditional HHE numbered report is the appropriate venue to provide this access, given the time taken to produce some of these reports, or whether more rapid access to appropriately vetted data and reports might provide a more effective and efficient alternative. Providing comprehensive, accessible data is now a requirement for large-scale National Institutes of Health-sponsored extramural research, and the online provision of questionnaires, methods descriptions, and workplace diagrams allows accessibility of a more complete data set than that available through peer-reviewed published sources. Peer-reviewed publication, however, is also a critically important feature of all scientific discovery including HHEs, and the HHE Program has been successful in producing such outcomes as a result of some of its investigations. The HHE Program could potentially increase the relevance (and impact) of its activities by determining whether data and information collected, or recommendations made, during the course of an HHE should be publicly released prior to the publication of its numbered reports.

The numbered and letter reports in the committee's sample demonstrate the depth of expertise available to the HHE Program. Expertise from other parts of NIOSH was included in the investigation of hazards such as noise and electromagnetic field exposures. A high level of interagency cooperation and interaction with a variety of stakeholders in developing outreach materials was displayed, and these interactions were often relevant for policy development within NIOSH and beyond. The quality and depth of interaction between HHE Program staff and stakeholders is evident in major examples described in Boxes 3-2 through 3-5, as well as for information developed by the HHE Program concerning latex exposure (NIOSH, 1997c, 1998d), publications related to carbon monoxide as a prevalent and lethal hazard in outdoor settings (NIOSH, 1996b; U.S. Coast Guard, 2001, 2008; U.S. National Park Service, 2005); and the enhancement of understanding of indoor air quality issues (EPA and NIOSH, 1998; Mendell et al., 2002; Kreiss, 2005), among other topics.

Review of Letter Reports

The HHE Program provided the committee with 10 letter reports written in response to full site visit investigations, and 5 examples of letter and fax responses provided when site visits were considered unwarranted. The letters included names and contact information of the program personnel responsible for generating the report, and telephone contact was freely offered. Two of the letter responses were apparently written using a template, indicating efficient use of resources. Each was appropriately adapted to the specific concerns of the complainant and included useful references to bolster the generalized response.

All the letter reports could benefit from a short list of recommendations in priority order, similar to those in numbered reports, so that clear and immediate information may be provided. Apart from this distinction, the letter reports reviewed were generally of as high a quality as the numbered reports and often included appropriate references and extensive reports of findings in tables and appendixes. The examples provided reflect a response to customer service needs for direct contact and timeliness. Seven of the 10 site visit letter responses indicated the date of the request. The time from request to final response ranged from 1 to 11 months, including four letters issued 5 or more months after the request.

One letter report appeared extremely helpful because it included contact information for local and state personnel who could address specific requestor concerns. In another report, a nonspecific recommendation for medical evaluation for personnel with indoor air quality complaints (NIOSH, 2007a) would have been more helpful if accompanied by a link to professional resources such as the Association of Occupational and Environmental Clinics. The letter reports related to IEQ issues (NIOSH, 2007d, for example) described but did not provide reference to an Institute of Medicine (IOM) report. A website link to the IOM report could have been provided.

Stakeholder Satisfaction and the Followback Program

To determine the level of satisfaction with HHE field investigations and reports, the committee reviewed the responses to followback questionnaires (NIOSH, 2007b, e). The HHE process and reports were generally highly valued and useful, and some recommendations were being implemented. Timeliness of reporting still needed improvement, and stakeholders sometimes raised concerns regarding the feasibility or technical accuracy of HHE recommendations. Some stakeholders said they did not understand why their HHE requests were not deemed appropriate for field investigations. A high percentage of followback questionnaire respondents indicated that HHE Program staff members were very professional and knowledgeable and that the HHEs were conducted in a professional manner. Nearly all stakeholders who provided information to this committee were impressed with the level and breadth of expertise demonstrated by HHE staff. Based on reports reviewed by this committee, HHE Program staff members are well aware of the literature regarding different hazards or potential hazards associated with the requests they are receiving.

Improvements could be made to avoid selection bias in followback program results by identifying similar facilities for which HHEs were not requested and by comparing health hazards before and after investigations. Follow-up health questionnaires, examinations, or exposure monitoring could show significant improvement. These exercises, however, require resources the HHE Program may not have. Nonetheless, the committee believes the followback program is an important tool for program process improvement and could be made more valuable with input from program evaluation expertise. Current strategic goals include performance measures that would expand the current followback questionnaire feedback loop to address program impact and improve customer satisfaction. Continued efforts to improve the timeliness of reports, improve written and telephone communication, and enhance dissemination efforts were listed by the HHE Program as measurable objectives to improve customer response (see Table 3-1). No information was relayed regarding the resources required to accomplish the objectives.

Quality of Recommendations

The committee assessed the quality of recommendations contained in HHE reports, and many were excellent. As one example, in response to an employee-generated request for an HHE, the HHE Program conducted an evaluation of a government-owned, contractor-operated hazardous waste "tank farm." A great deal of negotiation with workplace management was required to conduct this investigation, but the result was a rapidly produced, thoughtful, and thoroughly referenced report that provided an overarching recommendation embedded with very specific recommendations (NIOSH, 2004b). This approach created a framework for managing specific problems across a complex worksite with multiple layers of responsibility. It provided a model that synthesized both the problems and the approaches to solving them.

The committee found several reports and recommendations that do not exhibit the same level of quality. For example, in one reviewed report that was focused on musculoskeletal issues and recommendations, a variety of other potential safety hazards were identified, such as inadequately plumbed eyewashes and slippery surfaces. Only superficial recommendations to control these hazards were provided and did not include references for more detailed information (NIOSH, 2005a).

Another report, issued 18 months after the initial request, clearly described serious outdoor carbon monoxide exposures, but offered only generic recommendations for remediation despite the evidence of ongoing work by other scientists in NIOSH to develop engineering controls (NIOSH, 2004c).

The quality and utility of the HHE Program recommendations have been targeted for improvement during previous program reviews (NIOSH BSC, 1997, 2006; RTI, 1997). A systematic response on the part of the HHE Program to enhance the quality of recommendations does not appear to have been made. The mixed quality of the recommendations in the reports and letters reviewed by this committee suggests that quality control measures could be upgraded. At present, technical and policy review is conducted at the branch level within NIOSH: only rarely, and at the discretion of the branch chief, is other scientific quality review obtained. Given the understandable tension between timeliness and the inevitable delays that additional layers of review would entail, the committee recommends a sampling strategy of recent reports for review for scientific content and especially for accuracy of recommendations. The strategy would be used primarily as a quality improvement or training mechanism. Such reviews could be obtained from scientists elsewhere in NIOSH, from extramural scientists, from practitioners in labor and industry, and perhaps from OSHA or the Mine Safety and Health Administration (MSHA). Toxicologists, epidemiologists, engineers, and others from academic research institutions or from prioritized industry sectors might review reports both for quality and to identify emerging issues.

Recommendation: Ensure that recommendations in HHEs are relevant, feasible, effective, and clearly explained. Such steps may include

a. Explanation of the relevance, feasibility, and impact of each recommendation in the text of HHE reports.
b. Priority-setting among recommendations in all reports to indicate those requiring immediate action in the targeted workplace.
c. Debriefing in NIOSH after site visits and report dissemination for determination of relevance and impact on a systematic basis (potentially missed opportunities to identify emerging health hazards could also be identified).
d. Modification of the followback surveys for use in assessing the relevance, feasibility, and impact of recommendations.
e. Enhancement of internal quality assurance by development of a formal program that may include the external review of a sampling of recent reports and technical assistance letters for scientific content, report completeness, and appropriateness of recommendations.

EMERGING ISSUES

Finding 8: The HHE Program is an effective tool for uncovering emerging issues.

Emerging issues in occupational health include newly created hazards such as those arising from nanotechnology;[6] newly discovered hazards such as diacetyl flavoring; and known hazards in new or previously overlooked populations, such as silica exposure among roofers. Nanoparticles[7] are recognized as a potential health hazard and have been identified as an emerging issue by the National Academies Committee to Review the NIOSH Mining Safety and Health Research Program (NRC and IOM, 2007). As described in Chapter 2, OSHA has the lead in controlling well-understood hazards through legally enforceable standards. NIOSH has the responsibility to investigate new and emerging hazards for which standards do not exist or may be inadequate.

The HHE Program is an important mechanism within NIOSH for identifying and investigating emerging issues. The results of HHEs are similar to case reports in the medical literature; they are not always as definitive or as easily generalized to other workplaces as are the results of epidemiological research. But the HHE Program can take action on emerging issues in much less time, for much less money, and with much more flexibility to modify investigations in response to changing circumstances. One of the difficulties NIOSH and the HHE Program face in adapting the program to address emerging issues is that HHEs are done in response to requests, rather than being self-initiated. Nevertheless, NIOSH and the HHE Program could do more to track emerging issues and promote appropriate requests to address emerging issues.

A resource for stimulating HHE requests related to emerging hazards is the 26 state OSHA programs, all of which have enforcement and consultation resources, and all of which are members of the Occupational Safety and Health State Plan Association. To date, however, there is no evidence of interaction with this organization. The program could establish a stakeholder group or groups to assist in identifying exposure circumstances or types of workplaces that could be the object

[6]Nanotechnology, sometimes called *molecular manufacturing*, is a branch of engineering that deals with the design and manufacture of extremely small electronic circuits and mechanical devices built at the molecular scale. Nanotechnology is often discussed together with microelectromechanical systems (MEMS), a subject that usually includes nanotechnology but may also include technologies higher than the molecular level (http://whatis.techtarget.com/definition/0,,sid9_gci213444,00.html, accessed March 28, 2008).

[7]Nanoparticles are microscopic particles with an aerodynamic diameter of less than 100 nm. They are sometimes referred to as ultrafine particles (http://sis.nlm.nih.gov/enviro/iupacglossary/glossaryn.html, accessed August 6, 2008).

of HHE requests likely to have high relevance or impact. The NORA sector councils may serve this function.

Newly Identified Hazards

Work Organization Issues

Work organization describes how work is managed and organized. Issues include long or irregular hours, awkward or repetitive motion, job changes due to new technologies, and excessive or conflicting job demands. Twenty years ago, work organization issues were all but ignored; today, there is a growing recognition that work organization can have a major impact on worker health. Two of the most common problems are musculoskeletal disorders and stress-related conditions. Work organization issues can arise in any working population, but they are especially important in the healthcare and other service industries and in disaster response.

A number of HHEs have addressed work organization and its impact on health. They include HHEs of musculoskeletal disorders at the *Los Angeles Times* (NIOSH, 1990); of electronic headset noise exposure among transcribers at Kaiser-Permanente in California (NIOSH, 2005b); of respiratory, gastrointestinal, dermal, and stress-related disorders among members of the New Orleans Police Department after Hurricane Katrina (NIOSH, 2006e); and of respiratory disease and mental disorders among New York transit workers exposed in the 9/11 attack on the World Trade Center (NIOSH, 2005c). NIOSH has recognized that work organization issues are doubly relevant to its mission: even if they themselves do not cause death and disability, they may increase an afflicted worker's vulnerability to more traditional health and safety hazards. The HHE Program is instrumental in identifying and responding to emerging issues associated with work organization.

Chemical Hazards

To be more proactive, the HHE Program and NIOSH might consider certain classes of chemical compounds as emerging issues. For example, organic aldehydes such as acrolein are highly reactive compounds and widely used in certain industrial processes. Acrolein, used in the production of acrylic acid, has known toxic properties and may be carcinogenic (ATSDR, 2007). Quinones are another group of chemicals of potential concern. Occupational exposure to quinones may occur in the dye, textile, chemical, tanning, photographic processing, and cosmetic industries (OSHA, 2007). NIOSH and the HHE Program could determine the industries in which these and other potentially hazardous compounds are used and could determine whether existing permissible exposure limits (PELs) are adequate.

New Technologies

Few NIOSH HHEs have dealt with the hazards of major new technologies, and much more could be done in this area. For example, NIOSH has published only one HHE in the area of nanotechnology—an investigation of carbon fiber exposure in a University of Dayton research laboratory (NIOSH, 2006j). That investigation did not include medical evaluations and focused on traditional industrial hygiene controls for fine particulates. However, an increasing number of workers are exposed to nanomaterials, not only in laboratories, but in the manufacturing of products such as cosmetics. In the future, NIOSH will be challenged by a rapidly increasing use of nanotechnology in the workplace, and by other new technologies such as biologically engineered products and manufacturing methods. The HHE Program should be able to make use of the research in the respiratory program at NIOSH to address these issues.

Other emerging issues could result from the use of genetic engineering in the pharmaceutical industry that may generate unforeseen chemical and biological exposures. The HHE Program could identify these issues through NORA sectors, improved links with local and state health departments, and surveillance data (see below). Emerging issues could also be identified through rapid response to requests in this area and by determining where these products may be generated and contacting employers and employee organizations about HHE requests.

Known Hazards Affecting Underserved Populations and Small Businesses

It is an HHE Program priority to honor HHE requests. However, if the program is to be fully relevant, it should seek out emerging issues by ensuring that workers and employers know about the HHE Program and understand the value of filing a request. For the most part, this has not been done. The program has passively relied on requests from a variety of sources rather than actively seeking out worksites with new and emerging hazards. Such worksites could be identified through the occupational health literature, including international publications; engineering and trade association literature on new technology; databases of industrial chemicals and their uses; and the systematic use of state occupational health surveillance systems. A good first step might be the development of a systematic approach to facilitate identification of known hazards for which PELs may be inadequate or nonexistent and the identification of the workplaces where the hazards may be encountered. A rich source of ideas may be NIOSH's own files. It would be especially useful for

NIOSH and the HHE Program to maintain and systematically review a tickler file of odd and unexplained findings to be reviewed periodically for follow-up and further investigation.

Previous sections of this chapter have addressed the need for the HHE Program to better communicate with non-union workers and underserved populations. Emerging issues are particularly difficult to identify among these workers, specifically because of lack of communication with them. Novel means of outreach to these populations might include community organization and worker centers, as well as publicly funded health clinics.

Recommendation: Initiate formal periodic assessment of new and emerging hazards. The committee recommends the following steps:

a. Evolve from a program that passively receives requests to a proactive program that seeks opportunities for field investigations.
b. Develop systematic approaches to identify hazards where OSHA permissible exposure limits are inadequate or nonexistent, to identify unknown hazards, and to identify known hazards encountered under new circumstances.
c. Establish and periodically review a tickler file of inconclusive or unexpected evaluation results to determine whether new trends or problems may be emerging.
d. Periodically meet with intramural and extramural research scientists and stakeholders in government, academe, labor, and industry to discuss specific unresolved evaluations, to review aggregate findings, and to solicit input about new or emerging hazards or interventions. The HHE Program could establish one or more stakeholder groups to assist in identifying exposure circumstances or types of workplaces that could be the object of HHE requests likely to have high relevance and impact. The NORA sector councils may serve this function.

EMERGENCY RESPONSE

Finding 9: The HHE Program is uniquely qualified to formulate information needed to safeguard the workforce responding during and following a disaster. The program is well prepared to deploy during emergencies.

Emergency preparedness is one of the most relevant aspects of the HHE Program. The strength of the HHE Program in emergency response comes from the vast experience staff gain through day-to-day activities, particularly through engaging with employers, workers, and unions at workplaces throughout the country. Essential abilities are developed, such as improvisation and flexibility, which lend themselves to emergency response. As described in Chapter 2, the HHE Program has served an important role in the response to natural and manmade disasters. HHE personnel were essential to NIOSH's response to Hurricanes Katrina and Rita, largely because of their expertise in staging investigations under adverse conditions and writing recommendations in clear language without technical jargon. The NIOSH response was rapid, efficient, and competent. Within days, NIOSH produced and disseminated onsite guidance for emergency responders, medical personnel, and cleanup workers. According to a retired employee of the U.S. General Services Administration, Office of Governmentwide Policy, Office of Mail Communications, the HHE Program had a role that was both relevant and appropriate during the anthrax crisis, helping to ensure the safety and security of federal employees (Bender, 2008).

Since 9/11, the program has reasonably and appropriately focused on the all-hazards scenario—chemical, biological, radiological, nuclear, and high yield explosives events—working with other NIOSH programs responsible for information. The HHE Program works with OSHA in a joint operations capacity: the HHE Program provides advice and guidance to OSHA on health hazards and issues, while OSHA focuses primarily on safety.

Review of Outputs Related to Emergency Response

The committee reviewed two numbered reports related to emergency response submitted by the HHE Program (NIOSH, 2004d, 2006e). Both were considered extremely thorough. One report described multiple health hazards encountered by the New Orleans Police Department (NOPD), documented symptoms, and identified clear risk factors for a number of outcomes, including respiratory illness, posttraumatic stress disorder, and depression (NIOSH, 2006e). This report, produced within seven months, includes 48 references and clearly identified findings. It is a model for exploring emerging issues in disaster settings. At the time of this writing, the results from this report were in press in a peer-reviewed publication, and a follow-up field investigation had been requested by the NOPD. These outcomes demonstrated both the quality of the new knowledge developed and the relevance of the findings not only to the appreciative requesting entity but also to other public entities that may face similar disasters.

RELEVANCE SCORE

The program's strategic goals are appropriate, although some intermediate goals and performance measures could be made more ambitious if resources were available. The committee finds that the HHE Program responds well to HHE requests, and that HHEs play a key role in addressing widespread occupational health issues, such as musculoskeletal disorders, and emerging issues, such as lung disease from diacetyl-based flavorings. However, the committee believes the program could do a better job of eliciting a broader array of requests, especially from underserved populations and those exposed to new or newly recognized hazards. Rapid changes in the economy and workforce have affected the nature of HHE requests. The HHE Program has generally met these challenges with a judicious use of its resources, while remaining true to its mission. The program could do even better by making greater use of available surveillance data and by sharpening its focus on emerging issues.

The HHE Program has established a triage process to determine how to best meet the needs of each requestor within the limits of available resources. The triage process leads to improved efficiency, but it would benefit from increased transparency and more sophisticated site visit selection. Reports sent to the requestors and other interested parties and published on the NIOSH website are well written and relevant, reflecting a high level of expertise, and are supported by appropriate documentation. However, the quality and utility of some of the recommendations in the reports have not always met the same high standards. This could be corrected through a better quality assurance program. The timeliness of the reports has consistently been an issue in the past, but the HHE Program has shown some improvement in this area.

The 9/11 attacks and Hurricanes Katrina and Rita put needed emphasis on emergency response. The HHE Program has responded well to these emergencies and is uniquely qualified to develop better ways to protect the U.S. workforce during such disasters. NIOSH and the HHE Program are well prepared to deploy during emergencies.

After consideration of the criteria provided in the Framework Document (see Box 1-2), the committee assigns the HHE Program a score of 4 for relevance. The committee was asked to evaluate whether program activities are in priority areas and whether the program is engaged in relevant transfer activities. Although the committee makes multiple recommendations for improvement, it finds that program activities are in priority subject areas and that the program is engaged in appropriate transfer activities. A lower score, according to the scoring criteria, would indicate either a focus on lesser priorities or a lack of information dissemination.

Had the committee not been required to give an integer score, it might have rated the program between 4 and 5.

In another example, an HHE was performed for an employer who managed home care services through a California county office (10,000 home care workers, primarily representing workers from underserved populations) (NIOSH, 2004a). Representatives from the county office reported a successful partnering of the HHE Program with the Labor Occupational Health Program (LOHP) of the University of California, Berkeley (Ayala, 2008). LOHP developed follow-up training to assist in implementation of HHE recommendations, which also contributed to the creation of a Labor Management Committee that discussed health and safety, as well as other issues. In this example, the HHE Program took advantage of local resources to maximize the effectiveness of HHE recommendations.

A third example of HHE Program-state occupational health collaboration was provided in comments by an industrial hygienist working at the Massachusetts Department of Public Health Occupational Health Surveillance Program (OHSP), which receives NIOSH funding for its occupational asthma surveillance activities (Pechter, 2008).[8] Over time, three referrals for investigation of potentially hazardous exposures associated with work-related asthma were identified, and the industrial hygienist contacted the HHE Program to conduct investigations at these worksites. New-onset asthma associated with exposure to a particular compound (3-amino-5-mercapto-1,2,4-triazole) was identified (NIOSH, 2003b; Hnizdo et al., 2004). As noted by the stakeholder: "The identification of a new asthmagen is important, not only for the protection of workers currently exposed, but also to the process of scientific inquiry about respiratory sensitizers and asthma prevention" (Pechter, 2008). The HHE Program had previously identified asthma in flock workers at one company site (NIOSH, 1998a), while the OHSP request led to the identification of hazards in a second plant (NIOSH, 2006a; see Box 3-3 for a more detailed description of flock and related HHE Program activities).

---

[8]These comments reflect personal opinions and are not necessarily those of the Massachusetts Department of Public Health.

4

Impacts of the Health Hazard Evaluation Program: Reductions in Harm and Transfer of Knowledge

DEFINITION OF IMPACT

The committee is charged with evaluating the impact of the National Institute for Occupational Safety and Health (NIOSH) Health Hazard Evaluation (HHE) Program. Specifically, the committee is asked to determine the impact of the HHE Program on

1. Reducing worker risk and preventing occupational illness in investigated workplaces;
2. Transferring program-generated information to relevant employers and employees beyond the investigated workplaces;
3. NIOSH research and policy development programs; and
4. The activities of regulatory agencies, occupational safety and health professionals and organizations, state and local health agencies, and others in the occupational health community, as achieved by transferring program-generated hazard and prevention information.

This chapter is organized into seven major sections. This section defines impact and the context in which the program is being evaluated. The next four sections are analyses of each of the four types of impact described above, addressing both proximal and distal impacts of HHE Program activities, as well as program limitations. The committee then evaluates the impact of the HHE Program's emergency response activities. The final section of this chapter describes the committee score and rationale for program impact.

Ideally, a review of the HHE Program would yield quantitative evidence of the reduction of illness or death (earlier defined as "end outcomes") at specific investigated worksites or similar workplaces throughout the country. Such quantitative evidence of impact is generally unavailable. In the field of occupational health, however, there is solid evidence that exposure to certain chemical or physical agents causes illness, injury, or death. In general, a reduction in the level or frequency of exposures is expected to reduce the number of workers who develop disease. Elimination of exposures can reduce the number of cases of disease in those already exposed, prevent disease in new hires, or both. Based on this cause-and-effect relationship, attempts are made to reduce disease risk by reducing or eliminating exposures to various agents. Sometimes, based on reduced exposure, it is possible to make estimates of disease or deaths avoided.

Reductions in exposure at one workplace as a result of HHE Program recommendations can lead NIOSH or other agencies to take action to reduce exposures in similar workplaces nationwide. Information regarding exposure reduction measures can be provided to employees at a site where an HHE was conducted, and then to occupational health professionals, public health practitioners, and state and federal regulators elsewhere through reports, hazard alerts, and other publications. It is conceivable that NIOSH would have the data to estimate the number of lives saved and the reductions in diseases that can be direct or indirect results of HHEs conducted. For example, HHE recommendations describe the actions needed to reduce exposures of elevated chemical or physical hazards identified during an investigation. Follow-up with employers, and especially employees, could yield information about actions taken to reduce exposures. With sufficient evidence that harmful exposures have been reduced or eliminated (for example, by substitution of one chemical or process for another), it may be possible to develop quantitative estimates of occupational illness or death avoided.

Existing occupational health and safety data are insufficient to support robust analyses of impacts of the HHE Program. This is not to say that the HHE Program does not have impact, but that there are inadequacies in the reporting system. Because occupational disease and death statistics in particular are limited in the U.S. health data systems, other evidence is examined by the committee to estimate HHE impacts. The committee began this process by looking at the HHE Program's strategic goals and determining the number of HHEs conducted relevant to each strategic goal and by hazard type (Table 4-1). Starting with a table provided by the HHE Program (NIOSH, 2007b:Table 3-1), the committee used the program's online HHE search engine to identify related reports.[1] This simple exercise yields

[1]See http://www.cdc.gov/niosh/hhe/ (accessed July 9, 2008).

TABLE 4-1 Distribution of HHE Investigations by Strategic Goal and Hazard Type

Strategic Goal	Hazard Type	HHEs (numbered reports)
1. Prevent occupational illnesses through reduced exposure to workplace hazards	Biological	Biosolids (1) Brucellosis (3) Latex allergy (7) Tuberculosis (46)
	Chemical	Tertiary amines (3) Asphalt (37) Chlorinated compounds (28) Lead in construction (13)
	Physical	Musculoskeletal hazards (77) Noise (244)
	Mixed	Global health (0) Indoor environmental quality (200) Metalworking fluids (23)
2. Promote occupational safety and health research on emerging issues		Flock, respiratory (7) Flavorings (10) Severe acute respiratory syndrome (0) Silica in roofing (9) Carbon monoxide and houseboats (5) Surface wipe methods for chemical decontamination (1)
3. Protect health and safety of workers during public health emergencies		Anthrax (2) Irradiated mail (1) Natural disasters (0) Hurricane Katrina (3) World Trade Center (3)

SOURCE: NIOSH (2007b; HHE search engine [http://www.cdc.gov/niosh/hhe/, accessed July 9, 2008]).

somewhat different results with different search terms used for the same issues, but the exercise illustrates that the HHE Program likely has had impact in areas relevant to its strategic goals.[2] The HHE Program provided the committee with information on activities related to disease reduction, such as data sent to the Occupational Safety and Health Administration (OSHA) used for federal standard setting, which can be expected to result in lower exposures and, therefore, reductions in disease.

[2]The number of relevant HHEs identified by the search engine for a given search term varied over the course of committee deliberations.

1. HEALTH HAZARD EVALUATION PROGRAM IMPACT ON REDUCING WORKER RISK AND PREVENTING OCCUPATIONAL ILLNESS IN INVESTIGATED WORKSITES

Finding 1.1: HHEs generally have a positive impact on improvement of occupational health at investigated worksites.

There are many relevant examples of HHE reports that contain information useful to reasonably estimate reductions of health risk in the workplace. Numerous examples exist of workplace changes that have occurred as a result of HHEs. These include some fairly minor alterations in processes or in the use of control technologies or housekeeping procedures (for example, hazard-specific cleanup procedures) that reduce exposures and possibly result in a rapid reduction in illnesses. Table 4-2 illustrates several examples of impact-producing HHEs conducted from 1978 through 2006.

The HHE Program tracks data about HHE requests through a management system including, since 2005, the approximate number of people exposed to hazards (NIOSH, 2007f). HHE Program staff estimate that nearly 400,000 employees have been at risk at sites where HHEs were requested between fiscal years 2005 and 2008. This is based on information received from 87 percent of all requestors (information is missing for the remainder of the requests) (NIOSH, 2007g). The management tracking system could be improved with greater follow-up to obtain initially unreported data, and with better estimates of the exposed worker population in field-investigated sites.

A previous program evaluation by the Research Triangle Institute (RTI, 1997) recommended that the HHE Program systematically survey employers and employees in workplaces where HHEs, technical assistance, or consultation had been provided to learn whether HHEs satisfied customer needs, resulted in improved workplace health and safety, and identified emerging problems. As described earlier in this report, the HHE Program may conduct a followback survey at the completion of an HHE. One program staff person is responsible for followback survey activities. The survey response rate is about 60 percent (NIOSH, 2007b).

Among followback survey responders (289) who indicated they knew about the actions resulting from an HHE, 62 percent reported actions to implement HHE recommendations, and another 12 percent reported that actions were planned. Corrective actions taken were primarily housekeeping (85 percent), whereas personal protective equipment, engineering or administrative controls, and exposure monitoring varied from 50 to 75 percent of the remainder. In a survey of 68 respondents, 62 percent reported that employee health had improved (NIOSH, 2007b). This information, although not rigorously confirmed as representative, is important evidence in evaluating whether HHEs reduce risks in the workplace.

TABLE 4-2 Examples of HHEs Resulting in Wide Impacts (1978-2006)

Hazard	HHE Report[a]	Impact
Dibromochloropropane (DBCP)	• HHE-77-103-474 (Occidental Chemical Company, Lathrop, Ohio) • HETA-78-004-1511 (Shell Chemical Company, Axis, Alabama) • HETA-81-040-1315 (Dole Pineapple Corporation, Lanai, Hawaii) • HETA-81-162-1935 (Maui Land and Pineapple Inc., Kahuli, Hawaii)	• DBCP, once widely used as a nematocide, has been found to cause sterility among agricultural workers. • HHE Program data were used by OSHA to promulgate a standard in 1979 (U.S. Department of Labor, 2008a) to require reduced exposures.
Lead	337 HHEs were conducted between 1978 and 1995 in a wide array of industries	• HHE Program data were used by OSHA in promulgating lead standards (U.S. Department of Labor, 2008a, b). • For 25 years, HHEs provided information regarding exposures and control measures used by evaluated industries, OSHA consultation, and enforcement activities.
Silica, roofing tiles, construction	• HETA-2003-0209-3015 (Diversified Roofing Inc., Phoenix, Arizona) • HETA-2005-0032-2985 (Petersen-Dean Roofing Systems, Phoenix, Arizona) • HETA-2005-0031-3055 (C & C Roofing, Phoenix, Arizona) • HETA-2005-0030-2968 (Headlee Roofing, Mesa, Arizona)	NIOSH Publication 2006-110 (NIOSH, 2006h), based on four HHEs, describes hazard silicosis and how to protect workers. CPWR—The Center for Construction Research and Training (CPWR)[b] uses this publication in a curriculum to train 20,000 roofers (P. Stafford, CPWR Executive Director, personal communication, April 2008).

TABLE 4-2 Continued

Hazard	HHE Report[a]	Impact
Diacetyl	See Box 3-2 for HHE history 1985-2006	• Recognition of a pulmonary disease caused by diacetyl. • Publications in public health literature. • Data currently being used by the California OSHA to develop a standard to protect workers. • ConAgra, a major manufacturer, initiated pulmonary surveillance and industrial hygiene reviews to reduce exposure (J. E. Lockey, M.D., Professor, University of Cincinnati, written communication, February 18, 2008). • States alerted pulmonary physicians to diacetyl and bronchiolitis obliterans. • These HHEs caused NIOSH's pulmonary research unit to perform extensive research regarding bronchiolitis obliterans and focused international attention on this occupational hazard. • Voluntary removal of diacetyl from manufacturing processes in some facilities.
Silica flour	• HHE-78-104-107 (Tammsco Incorporated, Tamms, Illinois) • HHE-79-103-108 (Illinois Mineral Company, Elco, Illinois)	• Recommended exposure controls used in U.S. Department of Labor Mine Safety and Health Regulation, 1980 (U.S. Department of Labor, 2008c). • Health Effects of Synthetic Silica Particulates, a symposium sponsored by ASTM Committee E-34 on Occupational Health and Safety and the Industrial Health Foundation, American Society for Testing and Materials (ASTM), Benalmadena-Costa (Torremolinos), Spain, November 5-6, 1979. • NIOSH Current Intelligence Publication (NIOSH, 1981) recommended to Indian Silica Flour Industries by C. Rice, Deputy Director, and S. Clark, Director, Education and Research Center, Department of Environmental Health, University of Cincinnati. • HHE site recommended to Dr. A. El-Safty, Cairo University, Cairo, Egypt, 2008, by C. Rice.

continued

TABLE 4-2 Continued

Hazard	HHE Report[a]	Impact
Synthetic fibers, flock	See Box 3-3; 8 HHEs were conducted between 1972 and 2004: • HETA-96-0093-2685 (Microfibres Inc., Pawtucket, Rhode Island) • HETA-98-0212-2788 (Claremont Flock Corporation, Claremont, New Hampshire) • HETA-98-0238-2789 (Spectro Coating Corporation, Leominster, Massachusetts) • HETA-2004-0013-2990 (Hallmark Cards, Inc., Lawrence, Kansas) • HETA-2004-0186-3011 (Claremont Flock Corporation, Claremont, New Hampshire) • HHE-77-114-529 (The Standard Products Company, Lexington, Kentucky) • HHE-72-33-129 (Barker Greeting Card Company, Cincinnati, Ohio) • HHE-80-214-799 (M and B Metal Products, Inc., Leeds, Alabama)	High-hazard HHEs and NIOSH follow-up research are only warnings. OSHA does not have a regulation.

TABLE 4-2 Continued

Hazard	HHE Report[a]	Impact
Musculoskeletal disorders	Numerous HHEs from 1980 to 2007 (see Box 3-4)	The HHE Program identified hazards causing disorders of muscles, tendons, and joints in diverse industries, with 50 percent in manufacturing. These HHEs led to reviews and practice documents and have informed OSHA and other regulatory bodies in rule-making (for example, U.S. Department of Labor, 2000). The HHE Program's body of work on this topic has stimulated major research activities within and outside NIOSH on work-related musculoskeletal disorders. NIOSH has used these HHEs as bases of other investigations among underreported populations (farm workers).
Metalworking fluids (MWFs)	The HHE Program website indicated 19 HHEs were published between 1981 and 2006[c]	An array of industries use MWFs, from aircraft and automobile plants to missile and hydraulic plants. NIOSH estimates 1.2 million American workers are exposed (NIOSH, 1998b). Exposures are associated with a range of illnesses, from dermatitis to asthma and other pulmonary effects. A publication coauthored by HHE staff described an investigation and associated hypersensitivity pneumonitis (HP), a rare but severe condition, with MWF exposure (MMWR, 1996). Another HHE identified a severe outbreak of HP at a small plant and was instrumental in abating exposures at the facility and quelling the outbreak (NIOSH, 2002d). Both investigations were published in a short time in *Morbidity and Mortality Weekly Report.* HHEs have been important in defining the effects of MWFs and providing control technologies in hazardous workplaces where no OSHA standards or guidelines have been established to control exposures.
Latex allergy	Numerous HHEs from HHE Program website	NIOSH identified powdered latex gloves as the risk factor for latex allergy. Massive adoption of powder-free latex gloves followed.

[a]Specific HHE Program reports mentioned in this table can be accessed at the HHE Program website. See http://www.cdc.gov/niosh/hhe/ (accessed July 25, 2008).

[b]Formerly known as the Center to Protect Workers' Rights.

[c]See http://www.cdc.gov/niosh/hhe/HHEprogram.html (accessed July 25, 2008).

Recommendation: The followback surveys should be modified for use in assessing the relevance, feasibility, and impact of recommendations made by the HHE Program.

Finding 1.2: The number of HHE investigations completed in the last decade has declined dramatically. Therefore, fewer sites receive the positive benefits that accrue as a result of completed HHE investigations.

The HHE Program received an average of 372 requests for investigations per year between 1997 and 2006 compared with a historic average of 498 per year prior to 1997. The number of field investigations declined from 126 in 1997 to 58 in 2006 (NIOSH, 2007b). One reason for the decrease in field investigations was an increase in the proportion of requests concerning indoor environmental quality (IEQ) issues that could be addressed by telephone calls and letters. In the same period, invalid requests increased from 15 to 46 percent.

One can speculate that reductions over the past 10 years in both total and valid HHE requests have resulted from loss of jobs in the industrial sector, fewer health and safety personnel among union and non-union industries, or fewer hazards in workplaces. The number of full-time equivalents (FTEs) on staff dropped from 63 in 1998 to 52 in 2006 (NIOSH, 2007b). The budget for travel to perform HHEs remained essentially the same from 2000 to 2007 at the NIOSH Hazard Evaluations and Technical Assistance Branch (HETAB) (NIOSH, 2007b), indicating that less money is available for field investigations given increases in travel costs. Not only does the budget create problems for hiring adequate numbers of personnel, but the infrastructure (such as equipment and physical infrastructure) needed for HHE field investigations is capital intensive, potentially limiting the number of field investigations.

The reasons for the drop in field investigations are difficult for the committee to assess with confidence. However, the number of completed HHEs has been reduced. As a result, the potential impact of HHEs could also be reduced. The committee concludes that there likely are budgetary constraints in terms of personnel and equipment necessary to carry out HHEs given the decrease in FTEs and relatively flat budget between 2000 and 2007.

Finding 1.3: Large portions of the labor force—particularly those from traditionally underserved populations—are unaware of HHEs. Hence, they do not request the service and are unaware that a substantial body of HHE work exists to assist them.

One factor that may contribute to the decrease in HHE requests is that relatively fewer workers may be knowledgeable about the role HHEs can have in making their jobs safer. The structural shifts in the labor force away from unionized settings and toward the service sector, as well as the increase in contingent workers, have not been addressed by the HHE Program with additional outreach resources. The HHE Program lacks personnel to provide outreach to these workers who may benefit from HHEs. As an agency, NIOSH is strong in its abilities to conduct outreach when funds are available (for example, successful outreach to former nuclear workers funded by the U.S. Department of Energy through NIOSH; NIOSH and OERP, 2001). In response to the committee's request for stakeholder input (see Chapter 1 and Appendix D), many respondents indicated they did not know about the HHE Program. Others indicated they knew about the program and suggested outreach methods. At a December 2007 Worksafe Conference in California, several community and worker groups indicated no knowledge of what an HHE was or how to request one (Worksafe, 2007).[3] Recommendations from stakeholders about how to better communicate with these populations are included in the closing section of this chapter and in Appendix D.

A recent journal article by NIOSH investigators Cummings and Kreiss (2008) described the potential health risks of contingent workers, noting that there may be a lack of health and safety training and limited personal protective equipment for them. Contingent workers have nontraditional employment relationships; for example, they may be temporary or part-time workers or independent contract workers employed in refineries, agriculture, mining, or construction. The authors note that these workers are frequently young, female, African American, or Hispanic and have lower incomes and fewer benefits. This assessment was also made during a stakeholder presentation at the committee's second meeting (Gittleman, 2007). An exception to this characterization is construction workers who are members of the national building trades unions. These workers typically receive health and safety training and understand their rights.

During presentations to the committee, HHE Program staff acknowledged the need to give attention to shifts in the demographics and composition of the labor force. There may be sectors and employment groups not being served by HHEs that should be, and the committee suggests that day laborers, immigrant groups, some service industry workers, and low-wage workers may be examples. The HHE Program has had some success in reaching such underserved populations (see Box 1-1 for the definition of underserved populations). An HHE conducted for the Alameda County, California, Public Authority for In-Home Support Services, for example, led to successful and important HHE impact on a new kind of

[3]Worksafe is a California-based nonprofit organization established to promote occupational safety and health through education, training, technical and legal assistance, and advocacy. Members include labor and community groups, workers, occupational safety and health and other professionals, environmentalists, and others (http://www.worksafe.org, accessed March 30, 2008).

employment group—the home healthcare workers that the county office is serving (NIOSH, 2004a). The state contracts with the agency to use local independent contractors for home healthcare. Worksites are as varied as the clients needing home healthcare, and the workers are not in contact with one another. This particular HHE is also an excellent example of leveraging resources with state and local partners: a successful partnership among the federal, state, and local employment partners was necessary to conduct the HHEs, and innovative ways of communicating the results to multicultural workers working in multiple sites were required.

2. TRANSFER OF PROGRAM-GENERATED INFORMATION TO RELEVANT EMPLOYERS AND EMPLOYEES BEYOND INVESTIGATED WORKPLACES

Finding 2.1: The HHE Program disseminates its findings, sometimes widely.

Based on discussion with the HHE Program staff, program information is disseminated to employers and employees beyond the investigated workplace through

- HHE reports on the Internet accessed through the NIOSH website;
- CD-ROMs of HHEs available free to industries, workers, professors, and students;
- notification of subscribers by e-mail about various types of free documents;
- appropriate OSHA regional offices;
- scientific articles in technical journals;
- NIOSH documents such as Health Hazard Alerts released when a new body of information requires a special report to be brought to relevant communities (approximately one to five times per year);
- articles in trade publications;
- presentations at conferences and workshops; and
- alerts to other government agencies, including OSHA, at federal and state levels that may be in contact with employers and employees at similar workplaces.

Listed above are examples of mechanisms to transfer program-generated information to relevant employers and employees. Figure 4-1 is a flow diagram showing the potential transfer paths of HHE Program outputs to stakeholders and potential impacts they may have in general terms. HHE reports themselves, searchable on the NIOSH website, are important vehicles for affecting workplaces other than those investigated, given appropriate transfer mechanisms. CDs of HHE reports

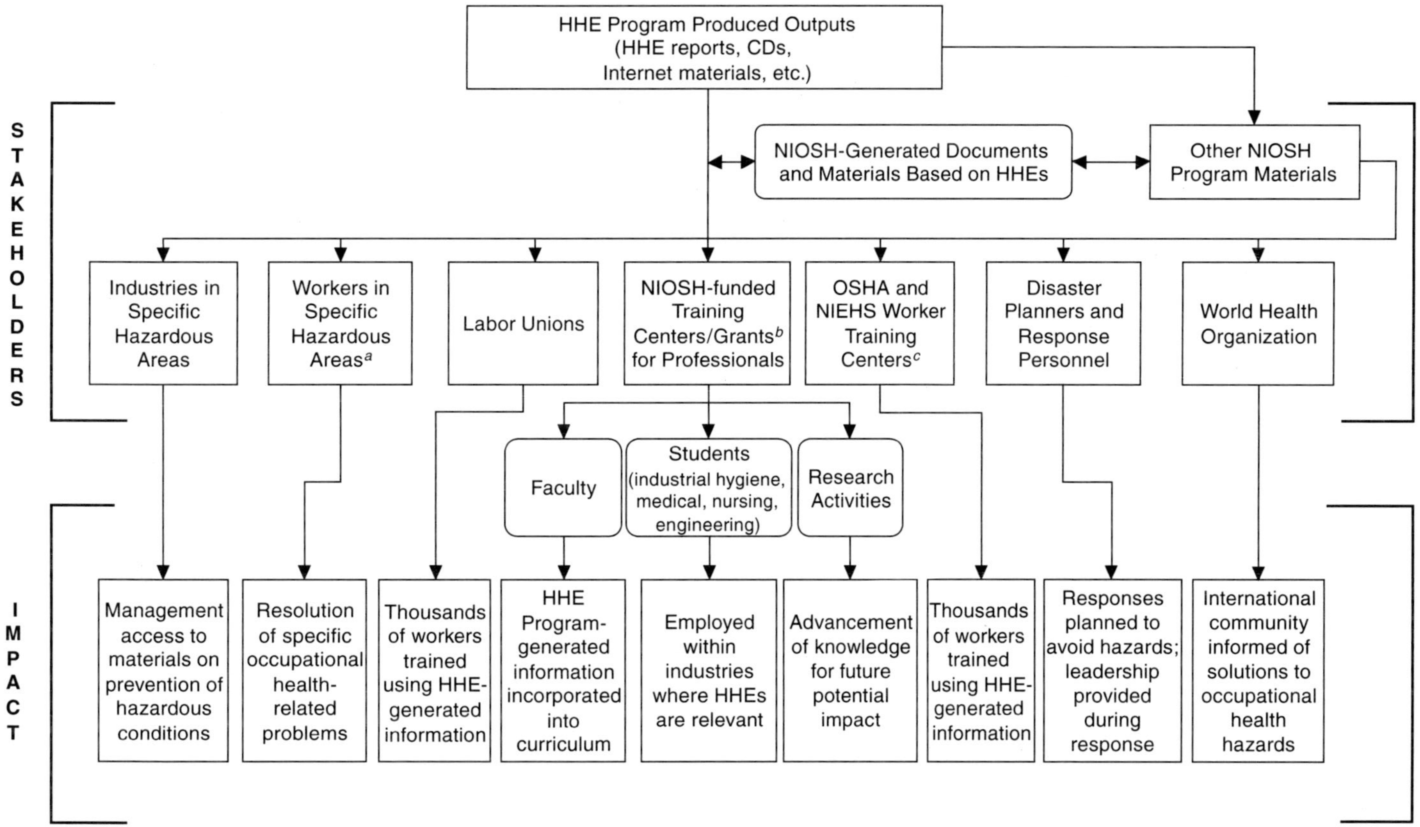

FIGURE 4-1 Routes of HHE Program impact beyond investigated workplaces.
NOTE: NIEHS = National Institute of Environmental Health Sciences.

[a] While the Internet is not currently accessed by all workers, more will have computer experience as the population ages, and this source of information will have broader impact.

[b] 17 training centers; 35 training grants.

[c] More than 50 OSHA and NIEHS Training Centers.

have been prepared, when funds were available, and provided free of charge to students in training in industrial hygiene, in occupational medicine and nursing, and as industrial safety specialists. Professionals in the field (including members of this evaluation committee) have also received such CDs. Members of the NIOSH Education and Research Centers (ERCs) training programs in master's and doctoral programs regularly use information collected in the conduct of HHEs (Carol Rice, Deputy Director, Education and Research Center for Occupational Safety and Health, University of Cincinnati, personal communication, April 2008). NIOSH has relationships with the International Labor Organization and the World Health Organization and shares hard copy and electronic documents for distribution to workplaces internationally (Paul Schulte, Director, Education and Information Division, NIOSH, personal communication, April 2008). The committee encourages relationships with these groups and with international NIOSH counterparts.

Because there are no U.S. Department of Labor regulations in place covering the use of many of the hazards identified in HHEs, these documents are valuable resources for employers, employees, consultants, and educators seeking to reduce illness and death in workplaces (see Table 4-2 for examples such as those related to metalworking fluids [MWFs] and musculoskeletal disorders). Annual reports, which might include information on the numbers and categories of HHE requests, industries and hazards involved, types of responses, resources required for responses, summaries of key findings from selected HHE reports, data on timeliness goals, and outcomes and impacts, might also prove useful tools for disseminating information about the program and its activities.

Finding 2.2: There is variable penetration of information into some communities.

Table 4-2 illustrates industries in which there has been widespread dissemination and transfer of information to relevant employers and employees. There is little to suggest, however, that HHE-generated information is received by employers and employees in small workplaces or by members of underserved populations. For this reason, the committee necessarily relied largely on anecdotal information. Below, the committee cites three examples of incomplete dissemination of HHE Program information.

Example 1: Trade Organizations

The committee heard from the vice president of manufacturing of a small flavor and fragrance manufacturing company in California (Speakman, 2008). An HHE was conducted at his facility to determine the cause and prevention of pulmonary disease among workers (NIOSH, 2007i).[4] The company representative had not heard of the flavoring-related illness prior to the investigation at his worksite. The committee noted that a relevant trade association was already aware of the hazard at the time the HHE was conducted. The representative informed the committee that he was not a member of the particular trade association from which he may have received relevant information but, because of cost, was a member of a different trade association. Discussion led the committee to conclude that reliance on trade associations as a means for the HHE Program to reach small business may not be sufficient.

After reviewing the list of HHE Program presentations made at trade conferences (NIOSH, 2007b:Appendix 2.14), the committee agreed that appearances in such venues are important and useful. However, the committee also received verbal and written stakeholder input that the HHE Program may need to increase its range of trade- and business-related venues. This is discussed in greater detail in Section 4 of this chapter.

Example 2: Small Residential Construction Companies

A representative of the CPWR—The Center for Construction Research and Training (CPWR)[5] noted that HHE Program success stories include HHEs that have resulted in better control technologies to reduce exposures to, for example, lead and silica (Gittleman, 2007). These HHEs were conducted primarily in union settings. Because of the mobile and contingent nature of the construction industry, however, many workers may still be exposed to risks that are well understood. The CPWR representative described a need to target high-risk residential construction by small companies that do not have the resources to investigate and solve their work-related safety and health problems.

Example 3: The Immigrant Labor Workforce

A stakeholder spoke to the committee on behalf of the Interfaith Worker Justice National Workers' Centers Network[6] and noted a general lack of knowledge and understanding among the immigrant population it represents of what the HHE Program does, how it differs from the Department of Homeland Security's

---

[4]The California Department of Public Health requested the HHE in 2006 with the company's cooperation, when two cases of pulmonary disease were identified at the facility. The HHE resulted in two additional cases of the illness being identified (NIOSH, 2007k).

[5]Formerly known as the Center to Protect Workers' Rights.

[6]Interfaith Worker Justice is a network of people of faith established to educate, organize, and mobilize the religious community to act to protect the rights of, and improve working conditions and benefits for, workers, especially low-wage workers (http://iwj.org, accessed June 1, 2008).

Immigration and Customs Enforcement (ICE) office, and how the HHE Program can benefit workers (Olíva, 2007). The inability of workers' centers to file HHE requests on behalf of workers, as well as the fear of government and negative experiences, indicates the need for special outreach to worker centers, community-based and faith-based institutions, and day labor organizations. The need for materials to be translated and for information to be transferred in culturally effective ways was also stressed by this individual.

In general, the committee finds evidence that HHE findings are disseminated widely to occupational health professionals, industries represented at technical meetings, certain trade associations, and a portion of the general public health community and worker training programs. The HHE Program stated in a written response to questions from the committee that it does not receive many requests from the agriculture and construction sectors. This was attributed to the temporary nature of the work, the mobility of the labor force, and the immigrant status of large parts of this labor force (NIOSH, 2007g:question #22). HHE Program staff members also indicated their awareness of the need to extend outreach, especially among the contingent workforce. The HHE Program has no formal mechanism to evaluate the effectiveness of knowledge transfer to employers and employees in facilities that have not been investigated.

Boxes 3-2 and 3-3 describe the occupational hazards of diacetyl and flock, respectively, and the HHE Program's essential role in determining the relationship between workers' exposure to these substances and serious respiratory diseases. However, the public and policy responses to these two substances have been completely different. There has been widespread media and regulatory attention focused on diacetyl, while flock has continued in production with relatively little attention. In some respects, the attention to diacetyl may be due to potential exposure among the general public as well as to workers. The HHEs related to diacetyl were an important part of a chain of events that will likely result in rule-making and have already led to limited voluntary substitution with other flavoring agents. In contrast, the HHEs related to flock exposure generated recommendations by NIOSH, but no attention at OSHA. The extent to which the flock industry follows the NIOSH guidelines is not known to the committee. It may be instructive to the HHE Program to assess the differences between the public responses to diacetyl and flock and to use these findings to improve information dissemination for similar issues in the future.

Recommendation: Develop a proactive, comprehensive information-transfer strategy for HHE Program outputs with better approaches to reaching wider audiences, including traditionally underserved populations. The HHE Program could

a. Use innovative techniques to reach small businesses and underserved populations, creating a broad array of mechanisms for communicating with diverse constituencies and attending to issues of literacy, language, and national-origin barriers.
b. Develop distribution mechanisms that are not Internet-dependent to complement Internet distributions.
c. Disseminate HHE results more broadly to groups likely to be affected, including distribution of HHE reports in the geographic regions where investigations are conducted.
d. Increase efforts to consolidate findings of multiple HHEs for specific hazards (for example, the compendia compiled for lead and other topics).
e. Develop improved methods of outreach to stakeholders so that workers and workplaces affected by new and emerging occupational health problems will be alerted quickly.
f. Supplement program outreach efforts by using community and small-business groups to translate HHE results and findings for their constituencies.
g. Leverage existing NIOSH, Centers for Disease Control and Prevention (CDC), and Department of Health and Human Services (HHS) resources to enhance technology transfer.
h. Evaluate, in a formal manner, the effectiveness of information-transfer programs, including knowledge transfer to employers and employees at worksites where HHEs have not been conducted.

3. IMPACT OF THE HEALTH HAZARD EVALUATION PROGRAM ON NIOSH RESEARCH AND POLICY DEVELOPMENT PROGRAMS

The HHE Program is a relatively small unit within NIOSH, which itself is a relatively small unit within CDC, which is a relatively small unit of HHS. A program of such small size and such a diverse mission needs to use the resources of its parent agencies to fulfill some of its functions. Furthermore, because the HHE Program itself does not have regulatory authority, its ability to have larger policy impacts is a function of its integration into the policy communication and decision-making structures of the agencies authorized to engage in policy development. In brief, the ability of the HHE Program to have impacts on other related research portfolios and policy development must in part flow through higher-level policy chains to have an impact on worker safety at the macro level (for example, beyond that of the individual HHE site level). This section of the report is divided into findings related to policy impact and findings related to impact on research programs.

Policy-Related Impact

Finding 3.1: The HHE Program does not have the authority to promulgate regulation but does inform the setting of regulations.

The HHE Program has had substantial impact on policy and the development of regulation to protect workers in several high-profile areas. Furthermore, in several high-profile areas, the HHE Program contributed knowledge that led to changes beyond those required by law or regulation on the part of industry. These changes have led to positive impacts on worker health and safety.

Policy is a multifaceted concept that is particularly complicated in the domain of public health. Federal policy-making and implementation are ultimately the responsibility of elected or appointed officials in the executive, legislative, and judicial branches. The HHE Program is part of HHS, a large and diverse executive cabinet agency, which is superordinate to CDC, which is superordinate to NIOSH, which in turn is superordinate to the HHE Program. As with all cabinet agencies, the policy agenda is ultimately set by the President of the United States. External political and organizational factors govern the types of policy impact that the HHE Program can have and are the dominant factors limiting the program's ability to have policy impact.

The HHE Program is quite small in both budgetary authority and personnel. Furthermore, its legislative mandate has been interpreted—through regulation—as very narrow with respect to the definition of what constitutes a valid request. Chapter 2 describes the regulatory interpretation. In this chapter, the committee interprets how the HHE Program leverages the resources of other occupational health actors to perform its mission and extend the impact of its activities and outputs.

The committee found substantial evidence that the HHE Program is positioned to provide high-quality technical advice—when requested—with some policy implications at the state level and for other parts of the federal government. For example, the HHE Program is represented on the National Occupational Research Agenda (NORA) sector councils and other federal occupational health committees with other agencies, such as OSHA, the Environmental Protection Agency (EPA), and the U.S. Navy (NIOSH, 2007b). In one case, the HHE Program teamed up with the Federal Highway Administration to evaluate asphalt exposure during paving (NIOSH, 2002a) and with the Transportation Security Administration to assess radiation exposure of baggage screeners (NIOSH, 2006i). In general, it would appear that most of such efforts are targeted at the federal level, but it is beyond the scope of the committee's charge to evaluate the impact of these mechanisms outside the HHE Program. Discussions with HHE Program staff during committee meetings suggest that staff hold a narrow view of the HHE Program role in policy development, either within or outside of NIOSH. The impression is one of purposefully keeping a low profile.

The committee has already noted the substantial evidence that supports the HHE Program's centrality in policy-relevant emerging hazard areas such as emergency response, and with hazards associated with diacetyl, MWFs, and latex gloves. During the course of this evaluation, the HHE Program became actively involved in a study of inflammatory neuropathy among swine slaughterhouse workers in Minnesota (MMWR, 2008). The committee considers this an example of how the HHE Program is positioned to respond quickly to emerging health hazards; at the beginning of this committee's deliberations, the issue was not of particular concern, but toward the end, the problem was nationally prominent. The findings emerging from HHE involvement are being used to develop safeguards in such facilities and may extend to other kinds of production involving high-pressure treatment of brain material. The committee concludes that the HHE Program has a very good reputation (among those who know about it) and that it is readily called upon in the event of new, emerging, and poorly understood problems.

At the time this report was being prepared, there was substantial evidence that HHE Program activity on the topic of diacetyl (a butter flavoring) has led to significant legislative efforts geared toward eliminating diacetyl from production processes. For example, California Assemblywoman Sally Lieber (D) introduced a bill to ban diacetyl use by 2009 (California Assembly, 2007).[7] In September 2007, the U.S. House of Representatives passed a bill to require OSHA to set a standard limiting diacetyl exposure (U.S. Congress, House of Representatives, House Education and Labor, 2007); as of June 2008, the bill was on the Senate Legislative Calendar. It is still not clear how the HHE Program participates in developing specific policy and research recommendations. For example, the committee was not able to determine exactly how the HHE Program originally became involved in the diacetyl issue. Diacetyl is used to flavor popcorn and a variety of other food products (see Box 3-2); however, it remains unclear how the HHE Program will expand knowledge about the hazards in the absence of HHE requests (for example, NIOSH, 2007i) or the very high profile cases related to popcorn manufacturing. This case perhaps highlights the difficulty the HHE Program has in proactively influencing policy development because it must be asked to conduct investigations or be called on by policy-makers for advice. The program's enabling regulations limit its ability to be proactive in participating in the regulatory process.

[7]This bill was approved by the California Assembly but was put into an inactive file of the California Senate in February 2008. As of June 4, 2008, no new action had been taken on this bill (California Assembly, 2008).

In an early case of high policy impact, in 1977, an HHE conducted in collaboration with the University of California developed evidence that dibromochloropropane (DBCP) caused sterility among workers at a facility in California (NIOSH, 1977). Several months later, OSHA used these program-generated data and experimental carcinogenesis data to regulate this chemical. This reduced the risk of sterility in people exposed to DBCP. Since DBCP is a component of pesticides, the impacts are far reaching for a number of different industries and worker groups. These efforts also contributed to the placement of this chemical on the EPA pesticide registry.

The body of MWF HHE reports had a significant input to the NIOSH *Criteria for a Recommended Standard: Occupational Exposure to Metalworking Fluids* (NIOSH, 1998b), which recommended the first authoritative exposure limit for MWFs. The findings led an MWF Standards Advisory Committee to recommend that an OSHA standard be set based on the NIOSH recommended exposure limit (REL), medical surveillance, and other provisions (Sheehan, 1999). OSHA declined to move forward with a standard, which is an external factor beyond the control of the HHE Program and NIOSH, limiting the impact of these findings. Nevertheless, the NIOSH REL, which is based in substantial part on these HHEs, has become the benchmark for equipment design and maintenance in significant industrial facilities. Committee members knowledgeable on this topic reported that new equipment manufactured in this industry complies with the NIOSH REL and that substantial progress has been made retrofitting older equipment to achieve this exposure limit.

The HHE Program played an important role in efforts to develop regulation related to indoor air quality. Ultimately, this policy process did not progress to an established regulation, but the centrality of HHE Program work in the deliberative process is indicative of its excellent reputation when higher levels of policy-making are introduced.

An OSHA employee who spoke with the committee suggested that prioritizing recommended controls in HHE reports would help OSHA to be more specific in standard setting (Kent, 2007). Overall, she highly commended the HHE Program, indicating that report findings are crucial for the regulatory process, but indicated that, regrettably, the HHE Program was not used to its fullest within OSHA or in industrial hygiene training communities. Ways of improving knowledge of the program are addressed in subsequent sections of this chapter and in Chapter 3.

The penetration of powder-free gloves into the healthcare market and practice is a direct result of HHEs conducted in healthcare delivery settings. Prior to HHE involvement, there was widespread use in the healthcare industry of powdered latex gloves, which caused a latex-related allergic reaction among some users. Several HHEs were conducted; as a body, these HHEs associated latex rubber protein with adverse allergic health reactions. The HHE investigators were among those who created the NIOSH Alert: *Preventing Allergic Reactions to Natural Rubber Latex in the Workplace* (NIOSH, 1997c). These were distributed very broadly within the healthcare delivery sector. Healthcare professional organizations cited the alert in making recommendations to provide alternatives to powdered latex gloves (e.g., Nagel, 1997 [American Chemical Society]; New York State Department of Health, 1998; Natural Rubber Latex Allergy discussion group, 2008).

Recommendation: Develop more extensive formal linkages and mechanisms with other parts of NIOSH, CDC, and HHS to enhance the capacity for involvement in policy-relevant impacts through

a. Promotion and increase in direct communication, especially with OSHA and state occupational safety and health agencies.
b. Alerts to NIOSH and CDC about HHEs that are relevant to policy-making outside the CDC system.
c. Continued regular use of the National Occupational Research Agenda (NORA) sector councils and the NIOSH Board of Scientific Counselors to disseminate information about the HHE Program.
d. Pursuit of a change in the HHE Program's legislative and regulatory authority to improve the capacity to identify hazards in need of HHEs, improve the ability to gain entrance to facilities when requested by treating physicians or community representatives, and address exposures other than chemical agents.

Finding 3.2: The HHE Program provides data and personnel to support NIOSH-recommended guidelines and NIOSH policy development activities.

The HHE Program communicates occasionally with other policy-related organizations, including those higher in the NIOSH hierarchy, to develop and forward policy recommendations. This activity may be limited by the organizational design of the agency (matrix management), the HHE Program's narrow interpretation of policy-relevant problems, and political concerns. The program has made contributions to NIOSH policy development through interactions with the NIOSH Authoritative Recommendations (AR) Program. According to written response to questions from the committee, the HHE Program has provided staff to support AR activities, including the preparation of written and oral testimony in support of proposed OSHA standards, and has provided data to support NIOSH policy development (NIOSH, 2008e). The HHE Program provided the committee information on 15 different areas in which the HHE Program has made or is making contributions toward NIOSH standards or policy developments (see Table 4-3).

The participation of the HHE Program in AR Program activities is evidence of how NIOSH fulfills its policy functions using HHE Program research. Examples of NIOSH policy functions include providing testimony to Congress, participating in OSHA regulatory processes, publishing guidance documents, and developing public statements for dissemination on the Internet. The HHE Program provides valuable support to these policy activities, which are outside the HHE Program regulatory mandate, but within that of NIOSH.

Finding 3.3: Relationships with certain state and local health departments appear to be strong.

The committee heard from stakeholders about important relationships between the HHE Program and certain local and state health departments, but the committee also learned of some state and local governments with which the HHE Program does not sufficiently communicate. The HHE Program has participated in NIOSH-sponsored surveillance meetings to reach state surveillance programs. It has made presentations to state-based organizations (for example, the Council of State and Territorial Epidemiologists) and has made contact with specific states. It has used the CDC Epidemic Intelligence Service (EIS) program to reach some state health departments and has trained state-based EIS officers at the Morgantown facility. It would not be expected that the HHE Program would work closely with the NIOSH Fatality Assessment and Control Evaluation (FACE) program because of FACE Program focus on acute traumatic fatalities as opposed to occupational illness.

Despite strong outreach efforts made in a general way and very good relationships developed in some specific cases (for example, with the California OSHA), outreach to state and local health departments is incomplete. Communicating with state health departments may not translate to communication with local health departments. Responses to the committee's request for stakeholder input indicate similar concerns about outreach to local health departments (see Appendix D for a summary of responses).

Research-Related Impact

In this section, the committee focuses on two aspects of research impact: those related to the development of laboratory or field-related experimental research programs at NIOSH and elsewhere, and those related to the impact on training the occupational health labor force. Section 4 of this chapter examines the impact

TABLE 4-3 HHE Program Interaction with the NIOSH Authoritative Recommendations Program

Subject Area	HHE Contribution
Tuberculosis	HHE investigators and data supported NIOSH testimony related to proposed OSHA rule on occupational tuberculosis
Ergonomics	• HHE investigators and data supported NIOSH testimony related to proposed OSHA rule on ergonomics (http://www.cdc.gov/niosh/docs/2001-108/pdfs/2001-108.pdf) • HHE investigators and data supported NIOSH guidance document (http://www.cdc.gov/niosh/97-117pd.html) • HHE investigators and data supported NIOSH science review document (http://www.cdc.gov/niosh/docs/97-141/)
Biosolids	HHE investigators and data supported NIOSH guidance document (http://www.cdc.gov/niosh/docs/2002-149/pdfs/2002-149.pdf)
Latex	HHE investigators and data supported NIOSH guidance document (http://www.cdc.gov/niosh/latexalt.html)
Indoor environmental quality	HHE investigators, drawing on their HHE field experience, helped develop NIOSH testimony to the U.S. Department of Labor on indoor air quality (Rosenstock, 1996)
Metalworking fluids	HHE investigators and data supported NIOSH policy document (http://www.cdc.gov/niosh/98-102.html)
Take-home lead	HHE investigators and data supported NIOSH policy document (http://www.cdc.gov/niosh/contamin.html)
Histoplasmosis	HHE investigators and data supported NIOSH guidance document (http://www.cdc.gov/niosh/docs/2005-109/)
Hexavalent chromium	HHE investigators and data supported NIOSH policy statement: NIOSH Comments on the OSHA Request for Information on Occupational Exposure to Hexavalent Chromium (http://www.cdc.gov/niosh/topics/hexchrom/pdfs/Cr(VI)_NIOSH_OSHA.pdf)
Respirator decision logic	HHE data supported NIOSH policy document (http://www.cdc.gov/niosh/docs/2005-100/appendix.html)
Hearing loss	HHE investigators supported NIOSH guidance document (http://www.cdc.gov/niosh/docs/96-110/pdfs/96-110.pdf)
Body art	HHE data supported NIOSH guidance (http://www.cdc.gov/niosh/topics/bbp/bodyart/)
Workers with disabilities	HHE investigators and HHE data helped develop NIOSH website content (http://www.cdc.gov/niosh/topics/wdd/default.html)
Emergency response	HHE investigators, drawing on their HHE field experience, helped develop NIOSH website content (http://www.cdc.gov/niosh/topics/emres/)
Skin notation	In progress

SOURCE: NIOSH (2008e).

of scientific publication specifically. Overall, for a program that is characterized as a service program, the committee found the HHE Program to be quite actively engaged in the applied scientific enterprise.

Finding 3.4: The HHE Program strongly influences research programs and the scientific body of knowledge.

The HHE Program has both indirect and direct influence on the research directions of NIOSH programs. For example, the various NORA committees develop research agendas for NIOSH programs they advise. According to a NORA sector committee member, some of the agendas are influenced by the work of the HHE Program (Barbara Silverstein, Research Director, Safety and Health Assessment and Research for Prevention, Washington State Department of Labor and Industries, personal communication, July 9, 2008). More directly, much of the research within, for example, the NIOSH Respiratory Diseases Research Program (RDRP), is based on unanticipated workplace hazards identified through the HHE Program. The HHE Program has stimulated research in areas such as the respiratory problems caused by flavorings, flock, waterproofing spray, and vaporized viruses. The output of this research is evidenced by the extensive publication list described in Section 4 of this chapter.

Aside from impacts already highlighted in this report, the HHE Program, since the 1990s, has contributed to other research areas, including the engineering control of noise exposure in indoor firing ranges; the etiologic relationships and health effect mechanisms associated with various musculoskeletal disorders and with male reproductive function associated with bicycle seats; and exposure monitoring techniques related to environmental tobacco smoke, nanotechnology, anthrax, and abrasive blasting (NIOSH, 2008e). Though it was beyond the committee's charge to review other NIOSH programs influenced by the HHE Program, the committee assumes, based on information received from the program, that it has influenced research within the NIOSH Cancer, Reproductive and Cardiovascular Diseases Program; the Engineering Controls Program; the Exposure Assessment Program; the Hearing Loss Prevention Program; the Musculoskeletal Disorders Program; the Nanotechnology Program; and the Personal Protective Technology Program, among others. The committee was not able to determine the level of influence based on the information presented. It is evident that HHE Program input has led to major contributions of new information in the literature and has increased the scientific knowledge base about workplace hazards.

One of the strengths of the HHE Program is its flexibility and responsiveness to a variety of occupational health problems. It maintains this strength with a very small budget and limited personnel, which requires the effective use of scientific resources located elsewhere. For example, NIOSH has an interagency agreement with the HHS National Toxicology Program (NTP) to support research on complex industrial exposures (Bucher, 2008). NTP evaluates chemical agents of public heath concern for chronic toxicity, including carcinogenicity and reproductive toxicity, using modern methods in toxicology and molecular biology (NTP, 2008). Chemicals to be tested may be nominated by anyone with a concern. In 1998, the HHE Program conducted two HHEs related to bromopropane, a group of chemicals for which there are no OSHA standards, at the request of the North Carolina Department of Occupational Safety and Health (NIOSH, 2002b, 2003d).[8] Comprehensive study of these chemicals would have required vast resources, and NIOSH subsequently nominated bromopropane in two chemical forms for evaluation by NTP. NTP evaluated bromopropane based on documented evidence of worker exposures (citing HHEs in its final reports) and the published evidence of reproductive and developmental toxicity, and defined the critical needs for data to improve assessment of toxicity to humans (NTP, 2002a, b).[9] HHE Program investigations led to two other published studies on bromopropane, one assessing the relationship between DNA damage and bromopropane exposure, and the other to better characterize the bromopropane exposure hazard and to evaluate the utility of a biomarker for assessing exposure (NIOSH, 2007b). NTP sometimes calls on NIOSH for its expertise to help evaluate the urgency of testing chemicals nominated by others for review. For example, the HHE Program may conduct HHEs and report back summary findings to NTP. Infrequently, NTP supports HHE activities, for example, an HHE related to cellulose insulation (NIOSH, 2001d; Bucher, 2008).

The HHE Program documented numerous productive arrangements with NIOSH laboratories for analytic purposes (NIOSH, 2007b).

Finding 3.5: The HHE Program has a significant impact on the training, development, and placement of program alumni inside and outside the HHE Program in the occupational health community.

A key output of the HHE Program is the occupational health professionals who have gone through training with the program. The HHE Program trains and places a large number of these professionals, constituting a significant transfer of human

[8]Bromopropane is used in spray adhesives; as a precision cleaner and degreaser; in the synthesis of pharmaceuticals, insecticides, quaternary ammonium compounds, and flavors and fragrances; and as a solvent for fats, waxes, or resins (NTP, 2002a, b).

[9]The NTP concluded that sufficient evidence existed to characterize 2-bromopropane as a reproductive hazard for humans (NTP, 2002b). 1-Bromopropane was determined to be a reproductive hazard for laboratory rodents, but insufficient evidence existed to conclusively show toxicity in humans (NTP, 2002a).

capital to the occupational health community through hosting international visitors, providing one- to three-month rotations or joint residency programs for occupational medicine residents, offering "tag-along" opportunities to participate in HHEs, supporting positions for EIS officers, and supporting summer positions for minority students to learn about occupational health (NIOSH, 2008d). HHE-trained professionals bring the skills, methodologies, and habits developed by the HHE Program to a broad array of public and private agencies. The committee finds that the HHE Program plays a significant role in the training of the occupational health labor force through its participation in such programs as the Commissioned Officer Student Training and Extern Program (COSTEP) and EIS. The committee is impressed by the commitment to occupational health fostered by participation in HHE Programs over the decades, attested to by committee members themselves and reflected in numerous written and oral statements made to the committee.

The occupational health community would benefit from continued HHE Program engagement in strategic recruitment of trainees, perhaps by partnering more extensively with ERCs and universities. Committee members conclude, based on their own experiences, that universities could and should be doing more to recruit the occupational health labor force of the future, and that the HHE Program could play a substantial role in the training and placement of that labor force. The funding of these programs is largely outside HHE Program control, which increases the need for the HHE Program to recruit actively from the CDC program training ranks.

There is limited tracking of program alumni (NIOSH, 2008d), and the committee finds little evidence that program alumni are used in any substantive capacity in advisory, case finding, or investigative roles (despite a high degree of enthusiasm and loyalty exhibited among program alumni encountered by the committee). The committee recommends the development of a program-level advisory board that would assist the HHE Program in leveraging its resources. In addition to program alumni, board members could include members of community groups and occupations not currently well served by the HHE Program. The advisory board could serve a recruiting and retention function, provide opportunities to use program alumni, assist with case finding, and provide expert advice about meeting some of the challenges facing the HHE Program given its limited resources and emerging challenges.

Recommendation: Use the HHE Program to develop occupational health professional resources. This could be accomplished through

a. **Increased recruitment of new investigators from universities, EIS, COSTEP, occupational medicine residencies, Education and Research Centers for Occupational Safety and Health (ERCs), and state**

and local health departments into HHE Program training rotations. This will require ongoing development of more attractive training, mentoring, and rotations.

b. **Tracking and mobilizing the extensive talent and commitment represented in the HHE Program-trained occupational health workforce. A network of HHE Program alumni could be fostered to help develop HHE opportunities. A program-level advisory board could assist the program in leveraging resources, serve a recruiting and retention function, assist in identifying emerging issues, and provide expert advice during normal program operations and when normal program operations are interrupted by emergency response activities.**
c. **Engagement and use of ERCs and other university-based training programs to involve trainees in HHE field investigations.**
d. **More formal collaborations with ERC faculty and other extramural researchers to assist with field investigation, dissemination, and training opportunities.**

4. TRANSFER OF PROGRAM-GENERATED HAZARD AND PREVENTION INFORMATION TO THE OCCUPATIONAL HEALTH COMMUNITY

The committee is charged with examining how the HHE Program transfers program-generated information to the occupational health community, with a particular focus on whether the information is presented appropriately. In this section, the committee begins by defining the occupational health community as a heterogeneous group of people requiring different strategies for effective transfer. The committee then turns to an evaluation of the specific knowledge transfer and exchange mechanisms used by the HHE Program to reach target audiences in the occupational health community.

The occupational health community is heterogeneous and includes scientists, occupational public health professionals, the public health labor force as a whole, companies, managers, members of the labor force, and institutions that represent individuals and groups. A comprehensive and appropriate technology transfer program would be targeted to the needs of this heterogeneous community. In order to assess this aspect of the HHE Program, the committee collected numerous examples of HHE communication strategies, solicited written comments from stakeholders, heard testimony from employers and employees affected by the HHE Program, and received detailed responses to related questions from HHE Program staff. Furthermore, a majority of the HHE Program performance measures (listed in Table 3-1) are directly related to transfer activities, including increased targeted marketing; increased number of links to the HHE Program website from external websites; greater use of eNews and Epi-X; publishing of compendia of findings, annual reports, and peer-reviewed publications; participation in interagency activities; and participation at trade meetings. Finally, recommendations related to transfer activities in prior evaluations of the HHE Program (NIOSH BSC, 1997, 2006; RTI, 1997) have been addressed by the program staff and are now represented in its current portfolio of knowledge transfer and exchange activities.

Examples of Transfer to the Occupational Health Community

Transfer Material Type: Numbered HHE reports

Finding 4.1: Numbered HHE reports are generally of high quality, of direct immediate benefit to investigated sites, and of benefit to the larger occupational health community.

As discussed in earlier chapters, the numbered HHE report is the primary output of the HHE Program, emanating directly from the enabling legislation and regulation. The committee reviewed numerous examples of numbered HHE reports, received input from report recipients, and solicited comments from the occupational health community about HHE reports. The committee also examined response patterns from the HHE followback survey program designed to assess satisfaction before, during, and after the conduct of an HHE investigation. HHE reports are generally of very high quality and useful to the specific site investigated and to the larger community of educated and informed occupational health professionals. The plain-language summary "Highlights" page included at the beginning of every numbered HHE report was specifically mentioned by stakeholders as being of particular use to workers.

Ongoing efforts in the development of the format of numbered HHE reports seem to increase both efficiency and effectiveness. When necessary, given the composition of the labor force, HHE reports are translated into other languages. The HHE Program is at heart an applied research program; however, the fact that some HHE reports have been translated to the peer-reviewed scientific literature is testimony to their high quality. In its request for input, the committee heard very positive feedback about the quality, efficiency, and usability of the HHE report process and products.

Transfer Material Type: Technical Assistance Letters

Finding 4.2: Technical assistance letters appear to be an efficient use of HHE Program resources when problems are well understood. Because

they are not disseminated widely, they are of little benefit to the larger occupational health community.

As discussed in Chapter 3, the committee reviewed 10 examples of technical assistance letters written by the HHE Program staff to individual requestors in response to valid HHE requests. HHE Program staff indicated that technical assistance letters were generated when a considerable amount of expertise on a topic already exists, enabling staff to provide detailed information to the requestors. For example, a letter response to a request concerning ventilation conditions in a health facility used the findings of a site-specific building inspection, a health inspection by the state authority, findings of a site safety committee, a report by the Institute of Medicine, a referral to relevant guidelines generated by a different state agency, national engineering standards, and a peer-reviewed publication (NIOSH, 2007d). This four-page letter provided very detailed guidance for remedying the problem, with appropriate resources provided as enclosures. Additional examples of technical assistance letters reviewed by the committee were on the topics of ventilation, exposures to well-defined chemical hazards, and exposure to biological hazards. In general, the letters were well documented. One of the letter reports pertaining to biological hazards included seven references to the peer-reviewed and regulatory literature (NIOSH, 2006f). Inclusion of such references in these reports effectively transfers knowledge of sources of relevant information. These letters do not appear to be disseminated widely, limiting the potential impact of the transfer of knowledge.

Transfer Material Type: Peer-Reviewed Publications

Finding 4.3: The HHE Program engages in a great deal of formal scientific publication related to its research.

The HHE Program provided the committee with 58 pages of titles of various papers disseminating program results dated from the early 1990s to the present. The titles include numbered HHE reports, journal articles, book chapters, NIOSH numbered documents, and other items appearing in trade journals.

From 1996 to 2007, the HHE Program issued well over 100 peer-reviewed publications, including those in major medical journals, such as the *Journal of the American Medical Association*, the *New England Journal of Medicine*, the *American Journal of Industrial Medicine*, the *Journal of Occupational and Environmental Medicine*, and the *American Journal of Surgical Pathology*. HHE Program scientists have also authored peer-reviewed articles in the leading general journal of public health, the *American Journal of Public Health*. Articles are similarly well placed in toxicology (*Neurological Toxicology, Toxicologist, Journal of Toxicology and Environmental*

Health, Critical Reviews in Toxicology); occupational health (*American Industrial Hygiene Association Journal, Applied Occupational Environmental Hygiene, Journal of Occupational and Environmental Hygiene*); and environmental health (*Journal of Environmental Monitoring, Environmental Health Perspectives*), among others. It is particularly noteworthy that HHE Program scientists have published 28 articles in *Morbidity and Mortality Weekly Report* (MMWR). Though MMWR is not peer reviewed, it has been a highly visible and respected publication in public health for many years (NIOSH, 2007b). These publications served to document advances in our knowledge of specific occupational illnesses and have had impacts on the development of treatment and intervention techniques to prevent or ameliorate the incidence of these illnesses.

Program transfer activities can be very thorough in some areas, and transfer occurs in a variety of ways. Since 1990, for example, there have been 39 numbered HHE reports, 20 journal articles, and 3 NIOSH numbered documents published—including proceedings of a workshop and a manual of analytical methods—all related to tuberculosis (NIOSH, 2007b). Additionally, seven items, such as chapters in books, and two items related to OSHA rule-making on occupational exposure to tuberculosis have been published. A more recent example is related to flavorings. Since 2002, 6 HHE reports, 10 journal articles (including in the *New England Journal of Medicine*), and 1 NIOSH numbered document have been published.

Transfer Material Type: Academic Presentations

> **Finding 4.4: The HHE Program presents research findings and staffs informational booths at major academic conferences related to occupational health.**

The committee reviewed an extensive 10-year record of regular presentations at major academic conferences, many of which were later published in the peer-reviewed literature. In addition to their substantial activities in the national occupational health community, HHE Program scientists have presented research at international conferences in Austria, Brazil, Canada, China, Finland, Germany, Italy, Mexico, Scotland, Sweden, South Africa, and South Korea. Examples of domains and specific academic-focused conferences include the following:

- General: American Industrial Hygiene Conference and Exposition; American Occupational Health Conference; World Congress on Safety and Health at Work; and American Statistical Association
- Medical: American Thoracic Society; American Academy of Allergy, Asthma and Immunology; Society of Diagnostic Medical Sonography

These presentations have been an effective mechanism for transfer of information from the NIOSH HHE Program to the public health community. The information communicated in the presentations can be expected to impact occupational health in a positive manner.

Transfer Material Type: Trade Presentations

Finding 4.5: There is evidence that the HHE Program publishes in trade journals and makes presentations to trade association meetings, but more can be done to reach a wider audience and a greater diversity of employer and industry types.

The committee reviewed a list of many presentations at trade conferences (NIOSH, 2007b). Examples of such presentation audiences include those

- in manufacturing: the American Automobile Manufacturers Association, the National Cotton Council of America, and the International Roofing Exposition;
- at professional associations: the National Hearing Conservation Association; and
- at regional meetings: the Kentucky-Tennessee Water Environment Association, the Phoenix Roofing Industry Silica Meeting, and the New England Biological Safety Association.

Appearances at such venues are important and useful. However, the committee received stakeholder input from invited speakers and in response to the committee's written request for input that the HHE Program may not be presenting at a wide enough range of trade- and business-related venues. In particular, it was noted that the HHE Program could do more to reach out to small-business associations. Furthermore, the characteristics of the construction industry and agricultural labor forces may require special outreach efforts. The associate director for Safety and Health Research at CPWR noted a need to target small, high-risk residential construction companies that do not have the resources to investigate and solve their work-related safety and health problems (Gittleman, 2007). It was noted that HHE Program success stories include lead and silica, but the speaker cautioned that the mobile and contingent nature of the construction industry may result in worker exposure to well-understood risks. The speaker suggested that additional venues likely to provide good opportunity for dissemination of program information are the annual NIOSH Build Safe Conference and the Chicago Safety Council Conference.

As described earlier in this chapter, the vice president of manufacturing for a small flavoring and fragrance company in California responded to questions during a committee meeting discussion (Speakman, 2008). Discussion led the committee to conclude that presentations made at trade association meetings are one important mechanism to transfer information, but many small businesses do not join their trade association or participate in association activities. Furthermore, the social dynamic at trade association meetings may make it difficult for business owners to have open dialogue about problems experienced with their production processes.

Transfer Activity Type: Interagency Cooperation

Finding 4.6: The HHE Program participates in a variety of intergovernmental activities that promote knowledge about occupational health matters.

The committee finds substantial evidence of engagement by the HHE Program with some parts of the federal government. Examples include involvement—sometimes over long periods—with single or ongoing working groups, such as the EPA Biosolids Exposure Measurement Workshop, the OSHA-NIOSH Interagency Health Outcomes, the Navy Occupational Health and Preventive Medicine Workshop, and the NORA Industry Sector Meeting.

Most such efforts appear targeted at the federal level. As discussed in Chapter 3, the committee heard from some state and local health officials who were unaware of HHE Program activities, including those occurring within their own jurisdictions. One e-mail respondent to the committee's request for input noted that the HHE Program was perhaps the "best kept secret in occupational safety and health" (Kiefer, 2007). A particularly poignant example of the program's "best-kept secret" status came from the commissioner of health of a major East Coast city. This individual had not heard of the HHE Program prior to being invited to speak to the committee, in spite of his former employment on the staff of U.S. Representative Henry Waxman (California), who is known for his activities related to health and healthcare reform. The idea was reinforced in a majority of the responses to the committee's written request for input (see Appendixes C and D). Although many respondents were quite familiar with the HHE Program, the most common suggestion for improvement was for the program to increase its outreach and visibility so that its services would be extended. HHE Program staff responded that the program has regular contact with the Council of State and Territorial Epidemiologists; the Epidemic Intelligence Service; state and federal OSHAs; and other parts of NIOSH and CDC (NIOSH, 2007h:response to question #21). The response further noted, however, that these efforts have not been formally evaluated. To the extent that this committee has conducted its own evaluation, it seems that the HHE Program has had significant involvement with some parts of the federal government and with some states. The evidence suggests that the HHE Program could enhance these efforts to encompass all potential federal and state partners. Furthermore, additional attention to local government agencies is also warranted.

Transfer Material: Compendia of Known Findings

Finding 4.7: The HHE Program has made progress in developing compendia of findings and recommendations about well-understood occupational health risks.

The HHE Program has developed four compendia of findings about occupational health risks associated with isocyanates, noise, tuberculosis, and lead (NIOSH, 2004e, 1998c, 2001b, and 2001a, respectively). The program is currently working on compendia on health risks faced by firefighters and those in the health-care sector.

The HHE Program tabulated the type and frequency of HHE recommendations, including 82 field investigations between 2000 and 2005 for which followback surveys were conducted (NIOSH, 2007f:Table G [a]). Of the reports, 65 percent recommended engineering controls, 33 percent recommended exposure monitoring, and 28 percent suggested medical surveillance. Although HHEs are often thought of as hazard-specific approaches to occupational health, the distribution of recommendation types suggests that the compilation of general process guidelines for use by employers, such as those already written for hazards including lead and tuberculosis (NIOSH, 2001a, b), may be an efficient use of program resources. For example, more than half of HHE reports recommended administrative controls (79 percent), and a large number of reports also recommended housekeeping controls (44 percent), labor-management communication (28 percent), and use of personal protective equipment (27 percent). If generalized process-oriented findings can be gleaned from the experience of the HHE Program across a variety of settings, then the program may consider compiling compendia of such findings.

Recommendation: Increase efforts to compile compendia of findings (such as those developed for isocyanates, noise, tuberculosis, and lead) when generalized process-oriented findings can be gleaned from the experience of the HHE Program in a variety of workplace settings.

Dissemination Strategies

Finding 4.8: A wide variety of dissemination mechanisms, including the Internet, are targeted to the professional occupational health workforce.

The committee has already noted significant program strengths in dissemination through peer-reviewed scientific literature and academic, professional, and trade association meetings. This section focuses on the ability of the HHE Program to disseminate findings to occupational health stakeholders beyond the professional labor force. The primary dissemination mechanism for HHE reports and other published materials is by e-mail and the Internet. Reports are posted on the HHE Program website and Epi-X, and are announced in eNews, a NIOSH e-mail newsletter. They are sent to, and made available through, the National Technical Information Service and federal OSHA regional offices. All of these mechanisms require (1) knowledge of the program and (2) a sophisticated ability to conduct an electronic search. The issue of whether people know about the HHE Program has already been discussed in detail. This section addresses the issue of an Internet-based dissemination strategy.

An Internet-based dissemination strategy does not serve people who do not have regular access to or knowledge about using the Internet. Such populations may include non-native language speakers, the working poor, and ethnic and racial minorities. The committee also notes that the program's Internet-based dissemination mechanism is difficult to maneuver, even for well-educated and well-informed healthcare professionals. Several respondents to the committee's request for input mentioned difficulty using the HHE Program website and emphasized the difficulty in finding HHE reports (see Appendix D for a summary of comments). Several written comments emphasized the difficulty of finding HHEs and also addressed the need for improvements to the search engine. An illustrative comment from an anonymous occupational and environmental physician and industrial hygienist in academe follows:

> The online search mechanism at the CDC website could be more user-friendly. In addition to the search by terms, an alphabetical categorization by main topic (exposure and/or disease) would be helpful. Also, the availability of HHE data could be better publicized. I have been aware of HHEs for years, but rarely do I hear about them from other sources.

Although the committee commends the HHE Program for making HHEs available online, committee members have had similar difficulty using the HHE search engine. In the preparation of Table 4-1, the committee noted apparent inconsistencies in the way keywords are assigned to HHEs. For example, in an effort to locate HHEs related to work-related musculoskeletal disorders in the program's online search engine (http://www.cdc.gov/niosh/hhe/), 77 matches were found when a search was conducted using the search term "musculoskeletal hazards," 113 matches with "musculoskeletal," and 92 matches with "ergonomic." Approximately 63 percent of the "ergonomic" results appeared within the results of the search under "musculoskeletal." Similar results were observed using search terms such as "indoor environmental quality," "IEQ," "indoor air quality," "respiratory indoor environmental quality," and "respiratory IEQ."

As early as the 1997 evaluation of the HHE Program by the Board of Scientific Counselors (NIOSH BSC, 1997), there were recommendations to improve linkage to extant occupational websites at OSHA, labor unions, and other interested stakeholders. Eleven years later, such measurable performance goals appear in the HHE Program strategic plan. Several respondents to this committee's request for input also noted the need for better linkage between the HHE Program and OSHA and other NIOSH websites. One respondent suggested that linkage between the HHE Program and local health departments, state occupational safety agencies, and industry associations could be improved, and an invited speaker to a committee meeting from OSHA Region 1 suggested that HHE Program success stories could be posted on the OSHA website (Kent, 2007). The committee encourages the HHE Program to work toward its performance measures of increasing the percentage of relevant websites with links to its own website (see Table 3-1).

The committee was told by HHE Program leadership that NIOSH is responsible for HHE webpages, editorial and graphics staff, and some administrative staff services. Although seeking economies of scale makes sense for a program as small as the HHE Program, it is also important that NIOSH provide adequate support to assist in the program mission. As revealed by its difficulty of use, it would seem that the HHE Program website is not a high priority for NIOSH.

The HHE Program appears aware of the limitations of reliance on the Internet for dissemination, as reflected in the program's strategic priorities and program descriptions of outreach activities. Respondents to the committee's request for input identified a number of ways in which dissemination could be expanded, especially to reach underserved populations. The responses are summarized in Appendix D.

In brief, the HHE Program appears aware of the need to diversify its social marketing efforts, and program staff informed the committee that it has engaged in pilot networking in Cincinnati, has conducted limited customer surveys, and is currently planning a targeted marketing campaign as part of its strategic plan. The recent employment of a full-time health communications expert will likely help move the program in the right direction. Participation in the NIOSH r2p initiative[10] may be promising, but depends greatly on the adequacy of funding and support for that effort, about which the committee collected no information.

Recommendation: Improve the searchability of the online HHE search engine by developing a list of standardized key words (an alphabetized listing of hazards and diseases would be beneficial).

Finding 5.1: The HHE Program is a recognized federal leader in its ability to respond quickly and effectively to new and emerging hazards, particularly in emergencies.

As is the case for other activities of the HHE Program, measuring the impact of its emergency response activities is difficult. Data related to impact do not exist, and impact in this area is more difficult to assess because the HHE Program does not serve as the lead agency during emergency response. HHE staff often work in concert with staff from other agencies, and often under unusual circumstances, making contributions by the HHE Program difficult to quantify. However, anecdotal evidence exists that can inform about the contributions of the program. As described in Chapters 2 and 3, HHE Program staff members are uniquely qualified to respond to emergencies and are often asked to assume leadership roles during emergency response. An important impact of the HHE Program, especially during emergency response, has been the subjective and effective validation of worker experiences and concerns. HHE Program procedures are established that allow staff to listen to workers in order to effect change and improve conditions.

Some noteworthy examples of the HHE Program's recognized excellence in emergency response are the prominent roles of staff in response to the World Trade Center disaster on September 11, 2001 (9/11). HHE staff participated in etiologic investigations of occupational exposures and the health effects among workers in a building close to the disaster site three months after 9/11 (Trout et al., 2002) and in later investigations of stress-related symptoms throughout the city (MMWR, 2002). The program also made important contributions in response to the 2001 anthrax attacks made through the U.S. Postal Service (USPS), highlighted in Box 4-1, and responding to issues related to the health effects of handling and opening mail irradiated to defend against biohazards (NIOSH, 2002c, e).

In response to natural disasters, the HHE Program conducted investigations of the mental and physical health of the New Orleans Police and Fire Department personnel following Hurricane Katrina (NIOSH, 2006d, e). As part of a CDC team,

[10] Research to Practice (r2p) is a NIOSH initiative focused on the transfer and translation of research findings, technologies, and information into highly effective prevention practices and products adopted in the workplace (http://www.cdc.gov/niosh/r2p/, accessed October 10, 2008).

BOX 4-1
HHE Program Response to Anthrax Contaminations

In October 2001, *Bacillus anthracis* (anthrax) spores were received in the mail by a news media office in Florida. HHE staff members were part of the response team that conducted the environmental evaluations of the affected building and of USPS offices where the mail was processed (NIOSH, 2007b). Ultimately, HHE investigators participated in 26 more emergency response investigations involving potential anthrax contamination in Washington, D.C., Florida, Connecticut, New Jersey, New York, Missouri, and Texas in a short period of time. Fifty-five separate sites were evaluated in New Jersey alone. HHE staff collaborated with researchers inside and outside NIOSH to design reliable standardized protocols for collecting samples and to use existing sampling technologies in innovative ways. Onsite technical assistance was delivered as part of the initial response to suspected biological threats (NIOSH, 2002b). NIOSH worked with other CDC offices, OSHA, and USPS to develop and disseminate guidelines on engineering controls and personal protective equipment for postal workers, mail handlers, and first responders (CDC, 2001; NIOSH, 2001c, 2002c), and helped design cleanup adequacy standards, implement decontamination procedures, and conduct post-cleanup assessments. NIOSH also trained Federal Bureau of Investigation and U.S. Coast Guard personnel, independent contractors, and others in appropriate anthrax decontamination procedures.

Information transfer was conducted by a variety of means. The emergency response teams communicated directly with teams at the investigated sites and helped staff the CDC command center, which provided 24-hour assistance for workers, employers, and the general public for an extended period of time. It is presumed that information disseminated through these mechanisms was based largely on the findings and expertise of HHE staff participating in the investigations. Additionally, HHE investigators authored or coauthored 17 peer-reviewed publications on the topic of anthrax; made 38 technical presentations; published 1 numbered HHE report and 3 web-based guidance documents; and testified to a congressional subcommittee regarding microbiological sampling methods for anthrax (NIOSH, 2007b). Within CDC, an educational video was produced for the benefit of postal workers (CDC, 2002).

Development of training tools and research in anthrax detection methodologies continued, stemming from the results of initial investigations. OSHA developed an illustrated and interactive web-based training tool on anthrax (OSHA, 2003). Sandia National Laboratories collaborated with Lawrence Livermore National Laboratories to develop a tool to assist cleanup personnel and officials at airports, other transportation centers, and high-traffic public buildings in the reoccupation of buildings following biological contamination, and tested the tool with the assistance of NIOSH staff (Sandia National Laboratories, 2005). The research for this tool was sponsored by the Department of Homeland Security and included partnerships with the San Francisco Bay area airports.

Reoccupation of facilities contaminated with anthrax and remediated and tested by investigative teams resulted in no new cases of the disease, indicating successful remediation efforts.

the HHE Program assisted local, state, and federal agencies in addressing occupational safety and health issues, performing surveillance and exposure assessments among workers, performing outreach to vulnerable workers, and disseminating occupational health data as the State of Louisiana and City of New Orleans rebuilt the city's public health system. The HHE Program also provided international assistance in the aftermath of the 2004 Indonesian tsunami that killed more than 225,000 people. HHE investigators coordinated and prioritized donations to speed the selection and manufacturing of personal protective and other equipment (NIOSH, 2007b).

The committee regards the information revealed through investigations conducted during the course of emergency response as useful and believes it will likely help responders in the event of similar future emergencies. Some emergencies, such as future anthrax or other biohazard attacks, may be avoided or their damaging effects mitigated, because of the guidelines and definitive recommendations established by NIOSH and the HHE Program. The U.S. Government Accountability Office (GAO) submitted a report to the House Subcommittee on National Security, Emerging Threats, and International Relations on agencies' need to validate sampling activities in order to increase the confidence in negative sampling results (GAO, 2005a). Although this report examined USPS, CDC, and EPA sampling methodologies, NIOSH provided guidance to USPS during its sampling and was consulted by GAO during preparation of the report.

Other NIOSH programs may have been informed by HHE Program activities, as exhibited by such publications as *Guidance for Filtration and Air-Cleaning Systems to Protect Building Environments from Airborne Chemical, Biological, or Radiological Attacks* (NIOSH, 2003c) and *NIOSH Interim Recommendations for the Cleaning and Remediation of Flood-Contaminated HVAC Systems: A Guide for Building Owners and Managers* (NIOSH, 2005d). The NIOSH National Personal Protective Technology Laboratory contracted with the RAND Science and Technology Policy Institute to review available databases related to disease, injury, and fatality data associated with personal protective technologies in emergency response (Houser et al., 2004).

Any evaluation of the impact of the HHE Program should consider that the program's resources (for example, funding, staff, scientific capacity) can and will be commandeered during a national emergency. The committee is not able to examine the effect of emergency response activities on day-to-day program activities in great detail because of the non-routine nature of emergency response. Financial repercussions of individual responses may differ depending on how the program is compensated for response activities. While emergency response may result in important positive impacts for those affected by the emergency, the program and

NIOSH should remain diligent to avoid negative impact on the protection of human health in more routine occupational settings.

Recommendation: Continue to provide guidance and recommendations during public health emergencies. To accomplish this, the HHE Program could

a. **Remain diligent by working with NIOSH management to avoid negative impact on routine activities of the HHE Program as a result of emergency response activities.**
b. **Develop a mechanism, such as the enlistment of help from training program participants and alumni, to ensure continuation of routine operations in the absence of staff involved in emergency response.**

IMPACT SCORE

As noted in the initial part of this chapter, it is difficult to find quantitative data to support an estimation of the degree of impact of the HHE Program in terms of decreases in numbers of workplace illnesses. Yet there is evidence that workers in investigated or similar workplaces have experienced reductions in exposures to health hazards and inappropriate work practices as a result of activities of the HHE Program. Because these reductions would be expected to lead to improved health for the workers, the scoring of impact of the HHE Program by the committee is based on intermediate outcomes, namely reductions in exposures to health hazards or an improvement in work practices.

The consensus of the committee is that the HHE Program can be highly effective in investigating hazards and providing advice to workplaces that make requests. The HHE Program is unique in the occupational health community in investigating unanticipated or underappreciated hazards in the workplace and relating them to the exposure or occupational circumstances of the worker. However, the committee is concerned that the program is not reaching all occupational groups, particularly employees of small businesses and underserved populations. There is also a concern that limited funding and obligations for emergency responses might dilute the effectiveness of the program and cause it to veer from more routine industrial hygiene reviews.

The transfer of information to workplaces other than those investigated and to other agencies in the occupational health community has been extensive; however, the transfer is incomplete in that many people interviewed by the committee were unaware of the HHE Program. Therefore, more emphasis needs to be placed on reaching out to the whole occupational health community.

The HHE Program is found to have a strong impact on the development of NIOSH-recommended occupational guidelines and on the support of NIOSH testimony for proposed OSHA rules. The HHE Program also contributes significantly to the advancement of scientific knowledge through its publications in the open literature and interactions with research programs within NIOSH and other government agencies. Training programs through the HHE Program are thought to offer an excellent opportunity for increasing expertise in occupational health, and such programs, as well as the use of alumni from these programs in emergencies, should be expanded.

Based on the scoring system developed by the Framework Committee (see Chapter 1, Box 1-1) the committee ranks the impact of the HHE Program as 4. If the committee had not been restricted to using integer scores, it would have scored the impact of the HHE Program as between 4 and 5.

5

Vision and Recommendations

In previous chapters, the committee evaluated the relevance and impact of the Health Hazard Evaluation (HHE) Program. In this chapter, the committee describes its vision of what the HHE Program should be. The description provides a picture of what the ideal program would look like, including elements of the existing HHE Program. The committee concludes the chapter by summarizing recommendations provided in Chapters 3 and 4 of the report.

VISION FOR THE HEALTH HAZARD EVALUATION PROGRAM

The ideal HHE Program, as envisioned by the committee, would respond promptly to requests for evaluation of the chemical, physical, and biological hazards within the HHE legislative scope, serving all workers, including underserved populations. The program would do what the National Institute for Occupational Safety and Health (NIOSH) can do uniquely—emphasize identification of health effects by combining medical investigation and industrial hygiene (exposure assessment) techniques with epidemiological and clinical toxicological perspectives, which may also involve development of new measurement and control techniques. The program would communicate results promptly to the target workplace, to workplaces with common exposures, and more broadly, to the public health community. In addition, the program would respond to requests from clinicians and community organizations with ties to workers, in addition to the current legislatively supported requests. After completion of investigations, staff would follow back on individual investigations to document and report abatement of hazards as well as follow back on grouped investigations to document abatement on a national basis. The ideal program would appeal to company and industry leaders, drawing them in as partners to prevent illness and injury industry-wide.

Ideally, the HHE Program would be at the cutting edge of identification of emerging occupational health hazards. This could be done through systematic use of scientific literature and surveillance data, including those generated by NIOSH, and effective knowledge management. In this way, the HHE Program could identify heretofore unrecognized hazards as well as known hazards for which permissible exposure limits (PELs) appear inadequate or for which an HHE would facilitate Occupational Safety and Health Administration (OSHA) standard setting when information is limited. The program would also work toward identifying new industries or workplaces where there are limited data on potential hazards and toward providing opportunities for application of other NIOSH resources, such as toxicology or control technology design responses, to emerging workplace issues.

HHEs would consider opportunities to confirm or deny health effects arising from occupational exposures. HHEs are opportunities to identify gaps in protection at target workplaces, and are indicators of possible problems at similar workplaces generally. HHEs are opportunities to describe problematic exposure circumstances that may be used as teaching examples. The program would be used to train field investigators, including NIOSH employees and those from partner agencies, academic institutions, and international counterparts.

Finally, the HHE Program would continue to provide guidance and recommendations during public health emergencies.

In contrast, the HHE Program would not provide what others can do, such as routine industrial hygiene measurements or standardized responses to common problems like indoor environmental quality. For such situations, the best use of HHE expertise would be to develop generalized protocols for response to common problems.

RECOMMENDATIONS

Within current legislative funding constraints, the HHE Program has managed to perform well. The HHE Program is a highly relevant program that has had and promises to continue having a valuable impact on improving worker health. The committee recommends the program be continued and, as possible, expanded to conduct more field investigations over a greater range of hazards, especially among underserved populations. In this section, the committee makes general recommendations for program improvement. The recommendations are organized into eight categories and presented in an order consistent with the HHE Program logic model. The chapter and page numbers where recommendations appear in the text are provided. The committee does not set priorities among its recommendations; it prefers encouraging implementation of any of the recommendations as resources are available. The committee recognizes that implementation of many of these recommendations will depend on the availability of resources, but at the request of NIOSH, it refrains from making recommendations regarding resource allocations. The next step for the HHE Program is to consider how to allocate resources.

Recommendation 1: Conduct regular assessments of performance measures to determine whether available resources allow more ambitious goals *(Chapter 3, page 50; see Table 3-1 for recommendations related to specific goals).*

Recommendation 2: Improve the mechanisms by which requests for HHEs are sought and prioritized to include a broader array of requests from a wider variety of requestors. The program could achieve this through

a. Systematic use of professional meetings, scientific conferences, scientific literature, and surveillance data, including those generated by NIOSH, to assist in prioritizing field investigations and recognizing emerging issues *(Chapter 3, page 59).*
b. Implementing, as part of the triage process, a formal technical assistance mechanism to help requestors to formulate valid HHE requests. In cases where an HHE is not appropriate or where resource limitations prohibit an investigation, technical assistance should include referral to more appropriate NIOSH divisions or government agencies *(Chapter 3, page 69).*
c. Development of an explicit, written process for classifying and prioritizing HHE requests. Priority should be based on the gravity of the potential harm, the number of employees potentially at risk at similar workplaces or using similar work processes, the urgency of the problem, the potential to assess health outcomes, and the possibility of identifying emerging issues. Potential impact on standards and policy should also enhance the priority of an HHE request in the triage process. Relationship of the HHE to current research may be considered, but should not be the only or primary factor. The process should provide guidance on weighting these varying factors *(Chapter 3, page 69).*
d. Better formalizing of the triage process, including the identification of needed expertise, and improving the transparency of the process

to HHE requestors, while maintaining flexibility and speed *(Chapter 3, page 70).*

e. Establishment of formal relationships with organizations representing underserved populations, small businesses, and their employees *(Chapter 3, page 66).*
f. Enhancing HHE Program outreach to OSHA national and regional offices and to state health and labor departments to better communicate the functions and activities of the HHE Program, increase cooperation with these agencies, and provide more complete and timely feedback *(Chapter 3, page 62).*

Recommendation 3: Ensure that recommendations in HHEs are relevant, feasible, effective, and clearly explained. Such steps may include

a. Explanation of the relevance, feasibility, and impact of each recommendation in the text of HHE reports *(Chapter 3, page 74).*
b. Priority-setting among recommendations in all reports to indicate those requiring immediate action in the targeted workplace *(Chapter 3, page 74).*
c. Debriefing in NIOSH after site visits and report dissemination for determination of relevance and impact on a systematic basis (potentially missed opportunities to identify emerging health hazards could also be identified) *(Chapter 3, page 74).*
d. Modification of the followback surveys for use in assessing the relevance, feasibility, and impact of recommendations *(Chapter 3, page 74; Chapter 4, page 90).*
e. Enhancement of internal quality assurance by development of a formal program that may include the external review of a sampling of recent reports and technical assistance letters for scientific content, report completeness, and appropriateness of recommendations *(Chapter 3, page 74).*

Recommendation 4: Use the HHE Program to develop occupational health professional resources. This could be accomplished through

a. Increased recruitment of new investigators from universities, the Epidemic Intelligence Service (EIS), the Commissioned Officer Student Training and Extern Program (COSTEP), occupational medicine residencies, Education and Research Centers for Occupational Safety and Health (ERCs), and state and local health departments into HHE Program training rotations. This will require ongoing development of more attractive training, mentoring, and rotations *(Chapter 4, page 106).*

b. Tracking and mobilizing the extensive talent and commitment represented in the HHE Program-trained occupational health workforce. A network of HHE Program alumni could be fostered to help develop HHE opportunities. A program-level advisory board could assist the program in leveraging resources, serve a recruiting and retention function, assist in identifying emerging issues, and provide expert advice during normal program operations and when normal program operations are interrupted by emergency response activities *(Chapter 4, page 107).*
c. Engagement and use of ERCs and other university-based training programs to involve trainees in HHE field investigations *(Chapter 4, page 107).*
d. More formal collaboration with ERC faculty and other extramural researchers to assist in field investigation, dissemination, and training opportunities *(Chapter 4, page 107).*

Recommendation 5: Develop a proactive, comprehensive information-transfer strategy for HHE Program outputs with better approaches to reaching wider audiences, including traditionally underserved populations. The HHE Program could

a. Use innovative techniques to reach small businesses and underserved populations, creating a broad array of mechanisms for communicating with diverse constituencies and attending to issues of literacy, language, and national-origin barriers. The effectiveness of applied outreach should be evaluated in a formal manner *(Chapter 3, page 66; Chapter 4, page 97).*
b. Improve the searchability of the online HHE search engine by developing a list of standardized key words (an alphabetized list of hazards and diseases would be beneficial) *(Chapter 4, page 116).*
c. Develop distribution mechanisms that are not Internet-dependent to complement Internet distributions *(Chapter 4, page 97).*
d. Disseminate HHE results more broadly to groups likely to be affected, including distribution of HHE reports in the geographic regions where investigations are conducted *(Chapter 4, page 97).*
e. Increase efforts to compile compendia of findings (such as those developed for isocyanates, noise, tuberculosis, and lead) when generalized process-oriented findings can be gleaned from the experience of the

HHE Program in a variety of workplace settings *(Chapter 4, pages 97 and 113).*

f. Develop improved methods of outreach to stakeholders so that workers and workplaces affected by new and emerging occupational health problems will be alerted quickly *(Chapter 4, page 97).*
g. Supplement program outreach efforts by using community and small-business groups to translate HHE results and findings for their constituencies *(Chapter 4, page 97).*
h. Leverage existing NIOSH, Centers for Disease Control and Prevention (CDC), and Department of Health and Human Services (HHS) resources to enhance technology transfer *(Chapter 4, page 97).*
i. Evaluate, in a formal manner, the effectiveness of information-transfer programs, including knowledge transfer to employers and employees at worksites where HHEs have not been conducted *(Chapter 4, page 97).*

Recommendation 6: Develop more extensive formal linkages and mechanisms with other parts of NIOSH, CDC, and HHS to enhance the capacity for involvement in policy-relevant impacts through

a. Promotion and increase in direct communication, especially with OSHA and state occupational safety and health agencies *(Chapter 4, page 101).*
b. Alerts to NIOSH and CDC about HHEs that are relevant to policy-making outside the CDC system *(Chapter 4, page 101).*
c. Continued regular use of the National Occupational Research Agenda (NORA) sector councils and the NIOSH Board of Scientific Counselors to disseminate information about the HHE Program *(Chapter 4, page 101).*
d. Pursuit of a change in the HHE Program's legislative and regulatory authority to improve the capacity to identify hazards in need of HHEs, improve the ability to gain entrance to facilities when requested by treating physicians or community representatives, and address exposures other than chemical agents *(Chapter 4, page 101).*

Recommendation 7: Initiate formal periodic assessment of new and emerging hazards. To accomplish this, the HHE Program could

a. Evolve from a program that passively receives requests to a proactive program that seeks opportunities for field investigations *(Chapter 3, page 78).*

b. Develop systematic approaches to identify hazards where OSHA permissible exposure limits are inadequate or nonexistent, to identify unknown hazards, and to identify known hazards encountered under new circumstances *(Chapter 3, page 78)*.
c. Establish and periodically review a tickler file of inconclusive or unexpected evaluation results to determine whether new trends or problems may be emerging *(Chapter 3, page 78)*.
d. Periodically meet with intramural and extramural research scientists and stakeholders in government, academe, labor, and industry to discuss specific unresolved evaluations, to review aggregate findings, and to solicit input about new or emerging hazards or interventions. The HHE Program could establish one or more stakeholder groups to assist in identifying exposure circumstances or types of workplaces that could be the object of HHE requests likely to have high relevance and impact. The NORA sector councils may serve this function *(Chapter 3, page 78)*.

Recommendation 8: Continue to provide guidance and recommendations during public health emergencies. To accomplish this, the HHE Program could

a. Remain diligent by working with NIOSH management to avoid negative impact on routine activities of the HHE Program as a result of emergency response activities *(Chapter 4, page 119)*.
b. Develop a mechanism, such as the enlistment of help from training program participants and alumni, to ensure continuation of routine operations in the absence of staff involved in emergency response *(Chapter 4, page 119)*.

References

ACGIH (American Conference of Governmental Industrial Hygienists). 1989. *Threshold limit values and biological exposure indices for 1988-1989.* Cincinnati, OH.

Associated Press. 2007 (September 6). Food companies plan to halt use of suspect butter flavor. *Boston Globe* (online). http://www.boston.com/business/globe/articles/2007/09/06/food_companies_plan_to_halt_use_of_suspect_butter_flavor/ (accessed October 10, 2008).

ATSDR (Agency for Toxic Substances and Disease Registry). 2007. Acrolein, CAS #107-02-8. Division of Toxicology and Environmental Medicine ToxFAQs™. Atlanta, GA: DHHS, Public Health Service. http://www.atsdr.cdc.gov/tfacts124.pdf (accessed July 23, 2008).

Ayala, L. 2008. Conference call with the Committee to Evaluate the NIOSH Health Hazard Evaluation Program, January 15, Irvine, CA.

Bender, J. 2008 (February 24). Personal response to questionnaire from the Committee to Review the NIOSH Health Hazard Evaluation Program.

Bucher, J. 2008 (July 15). Written communication from Associate Director, National Toxicology Program.

Buckle, P., and J. Devereau. 1999. Work-related neck and upper limb musculoskeletal disorders. Chapter 3.3. European Agency for Safety and Health at Work, Luxembourg. ISBN: 92-828-8174-1. http://osha.europa.eu/publications/reports/201 (accessd January 30, 2009).

California Assembly. 2007. An act to add Section 6404.7 to Labor Code, relating to occupational health. Assembly Bill No. 514, 2007-08 Leg., Reg. Sess. (Amended June 25, 2007). http://www.leginfo.ca.gov/pub/07-08/bill/asm/ab_0501-0550/ab_514_bill_20070625_amended_sen_v97.pdf (full text of bill) (accessed April 17, 2008).

CDC (Centers for Disease Control and Prevention). 2001 (October 31). CDC interim recommendations for protecting workers from exposure to *Bacilllus anthracis* in work sites where mail is handled or processed (updated from CDC Health Advisory 45 issued 10/24/01). Distributed via the Health Alert Network, DCDHAN-00051-01-10-31-ADV-N. http://www.cdc.gov/niosh/unp-mailrecs1.html (accessed July 11, 2008).

CDC. 2002 (January 9). Video: "Protecting your health" (for people who process, sort, and deliver the mail). Emergency Preparedness & Response. http://emergency.cdc.gov/agent/anthrax/training/mailhandlingvideo.asp (accessed July 11, 2008).

CNN.com/health. 2008 (February 28). Medical mystery solved in slaughterhouse. http://www.cnn.com/2008/HEALTH/conditions/02/28/medical.mystery/index.html?iref=newssearch (accessed March 3, 2008).

Cummings, K. J., and K. Kreiss. 2008. Contingent workers and contingent health: Risks of a modern economy. *Journal of the American Medical Association 299*(4):449-450.

EPA (Environmental Protection Agency) and NIOSH (National Institute for Occupational Safety and Health). 1998. Building air quality action plan. EPA Publication No. 402-K-98-001. Washington, DC: Author.

Eschenbacher, W. L., K. Kreiss, M. D. Lougheed, G. S. Pransky, B. Day, and R. M. Castellan. 1999. Nylon flock-associated interstitial lung disease: Clinical pathology workshop summary. *American Journal of Respiratory and Critical Care Medicine 159*(6):2003-2008. http://ajrccm.atsjournals.org/cgi/reprint/159/6/2003/ (accessed March 28, 2008).

GAO (U.S. Government Accountability Office). 2005a. Anthrax detection: Agencies need to validate sampling activities in order to increase confidence in negative results. Report to the Chairman, Subcommittee on National Security, Emerging Threats, and International Relations, House Committee on Government Reform, House of Representatives. GAO-05-251. http://www.terrorisminfo.mipt.org/pdf/gao05251.pdf (accessed July 10, 2008).

GAO. 2005b. Workplace safety and health: Safety in the meat and poultry industry, while improving, could be further strengthened. GAO-05-96. http://www.gao.gov/new.items/d0596.pdf (accessed March 28, 2008).

Gittleman, J. 2007. Personal response to questionnaire from the Committee to Review the NIOSH Health Hazard Evaluation Program.

Hnizdo, E., D. Sylvain, D. M. Lewis, E. Pechter, and K. Kreiss. 2004. New-onset asthma associated with exposure to 3-amino-5-mercapto-1,2,4-triazole. *Journal of Occupational and Environmental Medicine 46*:1246-1252.

Houser, A. N., B. A. Jackson, J. T. Bartis, and D. J. Peterson. 2004 (March). Emergency responder injuries and fatalities: An analysis of surveillance data. RAND Science and Technology Technical Report prepared for the National Personal Protective Technology Laboratory. TR-100-NIOSH. Santa Monica, CA: RAND. http://www.rand.org/pubs/technical_reports/2005/RAND_TR100.pdf (accessed July 10, 2008).

Kanwal, R., G. Kullman, C. Piacitelli, R. Boylstein, N. Sahakian, S. Martin, K. Fedan, and K. Kreiss. 2006. Evaluation of flavorings-related lung disease risk at six microwave popcorn plants. *Journal of Occupational and Environmental Medicine 48*(2):149-157.

Kent, M. 2007. Oral discussion with the Committee to Evaluate the NIOSH Health Hazard Evaluation Program, December 10, Washington, DC.

Kiefer, M. 2007 (November 27). Personal response to questionnaire from the Committee to Review the NIOSH Health Hazard Evaluation Program.

Kreiss, K. 2005. Building-related illness. P. 34 in *Preventing occupational diseases and injury, 2nd edition*, B. Levy, G. Wagner, K. Rest, and J. Weeks, eds. Washington, DC: American Public Health Association.

Last, J. M. (ed.). 2001. *A dictionary of epidemiology, 4th edition.* New York: Oxford University Press, Inc.

Lougheed, M. D., J. O. Roos, W. R. Waddell, and P. W. Munt. 1995. Desquamative interstitial pneumonitis and diffuse alveolar damage in textile workers: Potential role of microtoxins. *Chest 108*:1196-1200.

Massachusetts Office of Health and Human Services. 2008. Work-related asthma. Mass.Gov: For Consumers. http://www.mass.gov/?pageID=eohhs2subtopic&L=5&L0=Home&L1=Consumer&L2=Community+Health+and+Safety&L3=Workplace+Health+and+Safety&L4=Work-related+Asthma&sid=Eeohhs2/ (accessed July 21, 2008).

Materna, B. 2008. Oral discussion with the Committee to Evaluate the NIOSH Health Hazard Evaluation Program, January 15, Irvine, CA.

Mendell, M., W. Fisk, K. Kreiss, H. Levin, D. Alexander, W. Cain, J. Girman, C. Hines, P. Jensen, D. Milton, L. Rexroat, and K. Wallingford. 2002. Improving the health of workers in indoor environments: Priority research needs for a National Occupational Research Agenda. *American Journal of Public Health 92*(9):1430-1440.

Michaels, D., C. Morgan, and C. Monforton. 2008. A case of regulatory failure—Popcorn workers lung. http://defendingscience.org/case_studies/A-Case-of-Regulatory-Failure-Popcorn-Workers-Lung.cfm (accessed March 28, 2008).

MMWR (*Morbidity and Mortality Weekly Report*). 1996. Biopsy-confirmed hypersensitivity pneumonitis in automobile production workers exposed to metalworking fluids—Michigan, 1994-1995. *Morbidity and Mortality Weekly Report 45*(28):606-610. http://www.cdc.gov/mmwR/preview/mmwrhtml/00043265.htm (accessed July 28, 2008).

MMWR. 2001a. Update: Investigation of anthrax associated with intentional exposure and interim public health guidelines, October 2001. *Morbidity and Mortality Weekly Report 50*(41):889-893. http://www.cdc.gov/mmwr/PDF/wk/mm5041.pdf (accessed April 23, 2008).

MMWR. 2001b. Updated guidelines for evaluating public health surveillance systems: Recommendations from the Guidelines Working Group. *Morbidity and Mortality Weekly Report* 50(RR-13):2.

MMWR. 2002. Impact of September 11 attacks on workers in the vicinity of the World Trade Center—New York City. Reported by B. P. Bernard, S. L. Baron, C. A. Mueller, R. J. Driscoll, L. C. Tapp, K. M. Wallingford, and A. L. Tepper. *Morbidity and Mortality Weekly Report 51*(Special Issue):8-10. http://www.cdc.gov/mmwr/preview/mmwrhtml/mm51SPa3.htm (accessed July 10, 2008).

MMWR. 2008 (February 8). Investigation of progressive inflammatory neuropathy among swine slaughterhouse workers—Minnesota, 2007-2008. *Morbidity and Mortality Weekly Report 57*(5):122-124. On January 31, 2008, this report was posted as an MMWR Early Release on the MMWR website. http://www.cdc.gov/mmwr/preview/mmwrhtml/mm5705a3.htm (accessed March 28, 2008).

Nagel, M. C. 1997. A latex glove alert! Think twice before reaching for natural rubber PPE. *Chemical Health & Safety 4*(6):14-18. http://pubs.acs.org/hotartcl/chas/97/novdec/latex.html (accessed July 26, 2008).

Natural Rubber Latex Allergy discussion group. 2008. http://www.immune.com/rubber/index.html (accessed July 26, 2008).

New York State Department of Health. 1998. Preventing allergic reactions to natural rubber latex in the workplace: NIOSH Alert. http://www.nyhealth.gov/nysdoh/ems/latex.htm (accessed July 26, 2008).

NIOSH (National Institute for Occupational Safety and Health). 1977. Occidental Chemical Company, Lathrop. HHE Determination Report No. HHE-77-103-474. D. Whorton, T. H. Melby, H. A. Stubbs, C. R. Meyer, and T. Wilcox. Note: PDF link to report no longer viable at the NIOSH HHE website.

NIOSH. 1981. Silica flour: Silicosis (crystalline silica). *Current Intelligence Bulletin 36.* http://www.cdc.gov/niosh/81137_36.html (accessed June 25, 2008).

NIOSH. 1986. International Bakers Services, Inc., South Bend, Indiana. R. S. McConnell and R. W. Hartle. NIOSH Health Hazard Evaluation Report: HETA 85-171-1710. http://www.cdc.gov/niosh/hhe/reports/pdfs/1985-0171-1710.pdf (accessed March 28, 2008).

NIOSH. 1990. *Los Angeles Times*, Los Angeles, California. B. P. Bernard, L. J. Fine, and S. Sauter. NIOSH Health Hazard Evaluation Report: HETA-90-013-2277. http://www.cdc.gov/niosh/hhe/reports/pdfs/1990-0013-2277.pdf (accessed March 28, 2008).

NIOSH. 1994. Lundy Packing Company, Clinton, North Carolina. D. B. Trout and M. Kiefer. NIOSH Health Hazard Evaluation Report: HETA 93-0775-2398. http://www.cdc.gov/niosh/hhe/reports/pdfs/1993-0775-2398.pdf (accessed March 28, 2008).

NIOSH. 1996a. Microfibres, Inc., Pawtucket, RI. R. Washko, J. Burkhart, and C. Piacitelli. NIOSH Health Hazard Evaluation Report: HETA 96-0093-2685. http://www.cdc.gov/niosh/hhe/reports/pdfs/1996-0093-2685.pdf (accessed March 28, 2008).

NIOSH. 1996b. Safety and health topic: Carbon monoxide hazards from small gasoline powered engines. http://www.cdc.gov/niosh/topics/co/ (accessed July 22, 2008).

NIOSH. 1997a. Elements of ergonomics programs: A primer based on workplace evaluations of musculoskeletal disorders. A. L. Cohen, C. C. Gjessing, L. J. Fine, B. P. Bernard, and J. D. McGlothlin. NIOSH Publication No. 97-117. http://www.cdc.gov/niosh/docs/97-117/pdfs/97-117.pdf (accessed March 28, 2008).

NIOSH. 1997b. Musculoskeletal disorders and workplace factors: A critical review of epidemiologic evidence for work-related musculoskeletal disorders of the neck, upper extremity, and low back. B. P. Bernard, ed. NIOSH Publication No. 97-141. Cincinnati, OH: U.S. Department of Health and Human Services, Public Health Service, CDC, NIOSH.

NIOSH. 1997c. NIOSH alert: Preventing allergic reactions to natural rubber latex in the workplace. NIOSH Publication No. 97-135. http://www.cdc.gov/niosh/latexalt.html (accessed July 22, 2008).

NIOSH. 1998a. Claremont Flock Corporation, Claremont, New Hampshire. F. Daroowalla, M. L. Wang, C. Piacitelli, J. Burkhart, and W. Jones. NIOSH Health Hazard Evaluation Report: HETA 98-0212-2788. http://www.cdc.gov/niosh/hhe/reports/pdfs/1998-0212-2788.pdf (accessed March 28, 2008).

NIOSH. 1998b. Criteria for a recommended standard: Occupational exposure to metalworking fluids. NIOSH Publication No. 98-102. http://www.cdc.gov/niosh/98-102.html (accessed April 2, 2008).

NIOSH. 1998c. Health Hazard Evaluations: Noise and hearing loss, 1986-1997. Centers for Disease Control and Prevention. DHHS (NIOSH) Publication No. 99-106. http://www.cdc.gov/niosh/99-106pd.html (accessed October 27, 2008).

NIOSH. 1998d. Latex allergy: A prevention guide. NIOSH Publication No. 98-113. http://www.cdc.gov/niosh/98-113.html (accessed July 22, 2008).

NIOSH. 2000. NIOSH investigation of Gilster Mary Lee. A. Gomaa, G. Kullman, K. Fedan, P. Enright, P. Schleiff, P. E. Phillips, E. Simoes, and K. Kreiss. NIOSH Health Hazard Evaluation Report: HETA 2000-0401, technical assistance to Missouri Department of Health, interim report. http://www2a.cdc.gov/hhe/select.asp?PjtName=33418&bFlag=3 (accessed July 23, 2008).

NIOSH. 2001a. Health Hazard Evaluations: Issues related to occupational exposure to lead, 1994 to 1999. Centers for Disease Control and Prevention. DHHS (NIOSH) Publication No. 2001-113. http://www.cdc.gov/niosh/pdfs/2001-113.pdf (accessed July 26, 2008).

NIOSH. 2001b. Health Hazard Evaluations: Tuberculosis, 1990 to 1999. Centers for Disease Control and Prevention. DHHS (NIOSH) Publication No. 2001-116. http://www.cdc.gov/niosh/2001-116.html (accessed July 26, 2008).
NIOSH. 2001c. Interim recommendations for the selection and use of protective clothing and respirators against biological agents. DHHS (NIOSH) Publication No. 2002-109. http://www.cdc.gov/niosh/unp-intrecppe.htm (accessed July 11, 2008).
NIOSH. 2001d. NIOSH exposure assessment of cellulose insulation applicators. R. E. McCleery, J. McCullough, R. M. Hall, and J. E. Fernback. Health Hazard Evaluation Report: HETA-2000-03323-2827. http://www.cdc.gov/niosh/hhe/reports/pdfs/2000-0332-2827.pdf (accessed July 25, 2008).
NIOSH. 2001e. Simple solutions: Ergonomics for farmworkers. S. Baron, C. F. Estill, A. Steege, and N. Lalich. NIOSH Publication No. 2001-111. http://www.cdc.gov/niosh/pdfs/01-111.pdf (accessed March 28, 2008).
NIOSH. 2002a. Crumb-rubber modified asphalt paving: Occupational exposures and acute health effects. G. Burr, A. Tepper, A. Feng, L. Olsen, and A. Miller. NIOSH Health Hazard Evaluation Report: HETA 2001-0536-2864. http://www.cdc.gov/niosh/hhe/reports/pdfs/2001-0536-2864.pdf (accessed July 25, 2008).
NIOSH. 2002b. Custom Products, Inc., Mooresville, North Carolina. C. M. Reh, V. D. Mortimer, J. B. Nemhauser, and D. Trout. NIOSH Health Hazard Evaluation Report: HETA 98-0153-2883. http://www.cdc.gov/niosh/hhe/reports/pdfs/1998-0153-2883.pdf (accessed July 25, 2008).
NIOSH. 2002c. NIOSH fact sheet: Protecting workers from anthrax infection: Response from the National Institute for Occupational Safety and Health. NIOSH Fact Sheet. DHHS (NIOSH) Publication No. 2002-142. http://www.cdc.gov/niosh/02-142.html and http://www.cdc.gov/niosh/pdfs/02-142.pdf (accessed July 11, 2008).
NIOSH. 2002d. TRW Automotive, Mt. Vernon, Ohio. D. Trout and J. Harney. NIOSH Health Hazard Evaluation Report: HETA 2001-0303-2893. http://www.cdc.gov/niosh/hhe/reports/pdfs/2001-0303-2893.pdf (accessed July 28, 2008).
NIOSH. 2002e. United States Senate and House of Representatives, Washington, D.C. R. Hall, J. Hess, B. Bernard, M. Kiefer, J. Harney, D. Mattorano, R. McCleery, L. Delaney, M. Gillen, and K. Mead. NIOSH Health Hazard Evaluation Report: HETA 2002-0136-2880. http://www.cdc.gov/niosh/hhe/reports/pdfs/2002-0136-2880.pdf (accessed July 10, 2008).
NIOSH. 2003a. Bil-Mar Foods, Inc., Storm Lake, Iowa. B. S. King and E. Page. NIOSH Health Hazard Evaluation Report: HETA 2002-0257-2916. http://www.cdc.gov/niosh/hhe/reports/pdfs/2002-0257-2916.pdf (accessed July 25, 2008).
NIOSH. 2003b. ChemDesign Corporation, Fitchburg, Massachusetts. E. Hnizdo and D. Sylvain. Health Hazard Evaluation Report: HETA 2000-0096-2876. http://www.cdc.gov/niosh/hhe/reports/pdfs/2000-0096-2876.pdf (accessed March 28, 2008).
NIOSH. 2003c. Guidance for filtration and air-cleaning systems to protect building environments from airborne chemical, biological, or radiological attacks. DHHS (NIOSH) Publication No. 2003-136. http://www.cdc.gov/niosh/docs/2003-136/pdfs/2003-136.pdf (accessed July 10, 2008).
NIOSH. 2003d. Marx Industries, Inc., Sawmills, North Carolina. J. M. Harney, J. B. Nemhauser, C. M. Reh, D. Trout, and S. Schrader. NIOSH Health Hazard Evaluation Report: HETA 99-0260-2906. http://www.cdc.gov/niosh/hhe/reports/pdfs/1999-0260-2906.pdf (accessed July 25, 2008).
NIOSH. 2003e. Preventing lung disease in workers who use or make flavorings. R. Kanwal, G. Kullman, K. Kreiss, R. Castellan, J. Burkhart, K. Hilsbos, M. Akpinar-Elci, C. Piacitelli, and R. Boylstein. NIOSH Publication No. 204-110. http://www.cdc.gov/niosh/docs/2004-110/pdfs/2004-110.pdf (accessed March 28, 2008).

NIOSH. 2004a. Alameda County Public Authority for In-Home Support Services, Alameda, California. S. Baron and D. J. Habes. NIOSH Health Hazard Evaluation Report: HETA 2001-0139-2930. http://www.cdc.gov/niosh/hhe/reports/pdfs/2001-0139-2930.pdf (accessed March 21, 2008).

NIOSH. 2004b. CH_2M Hill Hanford Group, Inc. and United States Department of Energy, Office of River Protection, Richland, Washington. Y. Boudreau, J. Cardarelli, and G. Burr. NIOSH Health Hazard Evaluation Report: HETA 2004-0145-2941. http://www.cdc.gov/niosh/hhe/reports/pdfs/2004-0145-2941.pdf (accessed March 28, 2008).

NIOSH. 2004c. Lake Havasu Municipal Employees, Lake Havasu City, Arizona. R. McCleery, L. Tapp, J. McCammon, and K. L. Dunn. NIOSH Health Hazard Evaluation Report: HETA 2002-0393-2938. http://www.cdc.gov/niosh/hhe/reports/pdfs/2002-0393-2928.pdf (accessed July 22, 2008).

NIOSH. 2004d. NIOSH evaluation of air sampling methodologies for *Bacillus anthracis* in a United States Postal Service processing and distribution center, Trenton, New Jersey. R. E. McCleery, K. R. Martinez, G. A. Burr, and D. A. Mattorano. NIOSH Health Hazard Evaluation Report: HETA 2002-0109-2927. http://www.cdc.gov/niosh/hhe/reports/pdfs/2002-0109-2927.pdf (accessed July 23, 2008).

NIOSH. 2004e. A summary of health hazard evaluations: Issues related to occupational exposure to isocyanates, 1989 to 2002. Centers for Disease Control and Prevention. DHHS (NIOSH) Publication No. 2004-116116. http://www.cdc.gov/niosh/docs/2004-116/pdfs/2004-116.pdf (accessed October 27, 2008).

NIOSH. 2005a. Good Humor-Breyers Ice Cream, Hagerstown, Maryland. D. Habes and R. Driscoll. NIOSH Health Hazard Evaluation Report: HETA 2004-0117-2964. http://www.cdc.gov/niosh/hhe/reports/pdfs/2004-0117-2964.pdf (accessed July 22, 2008).

NIOSH. 2005b. Kaiser Permanente, Santa Teresa, Redwood City, and Santa Clara, California. R. L. Tubbs and C. A. Kardous. NIOSH Health Hazard Evaluation Report: HETA 2003-0280; 2003-0287-2974. http://www.cdc.gov/niosh/hhe/reports/pdfs/2003-0273-2974.pdf (accessed March 21, 2008).

NIOSH. 2005c. Metropolitan Transit Authority of New York City, New York City, New York. L. Tapp, K. Wallingford, C. Mueller, and S. Baron. NIOSH Health Hazard Evaluation Report: HETA 2002-0095-2955. http://www.cdc.gov/niosh/hhe/reports/pdfs/2002-0095-2955.pdf (accessed March 28, 2008).

NIOSH. 2005d. NIOSH interim recommendations for the cleaning and remediation of flood-contaminated HVAC systems: A guide for building owners and managers. http://www.cdc.gov/niosh/topics/flood/Cleaning-Flood-HVAC.html; http://www.cdc.gov/niosh/topics/flood/pdfs/Cleaning-Flood-HVAC.pdf (accessed July 14, 2008).

NIOSH. 2005e. NIOSH letter report: HETA 2004-0311 (dated January 3). M. M. Kawamoto. Provided in NIOSH response #1 to questions for the NIOSH HHE Program from the Committee to Evaluate the NIOSH Health Hazard Evaluation Program. Response to question #23. Cincinnati, OH.

NIOSH. 2005f. U.S. Magnesium, Rowley, Utah. E. A. Kim, B. P. Bernard, and E. J. Esswein. NIOSH Health Hazard Evaluation Report: HETA 2004-0169-2982. http://www.cdc.gov/niosh/hhe/reports/pdfs/2004-0169-2982.pdf (accessed March 21, 2008).

NIOSH. 2006a. Claremont Flock Corporation, Leominster, Massachusetts. V. Antao and C. Piacitelli. NIOSH Health Hazard Evaluation Report: HETA 2004-0186-3011. http://www.cdc.gov/niosh/hhe/reports/pdfs/2004-0186-3011.pdf (accessed March 28, 2008).

NIOSH. 2006b. Glister-Mary Lee Corporation, Jasper, Missouri. R. Kanwal, G. Kullman, K. Fedan, and K. Kreiss. NIOSH Health Hazard Evaluation Report: HETA 2000-0401-2991. http://www.cdc.gov/niosh/hhe/reports/pdfs/2000-0401-2991.pdf (accessed March 28, 2008).

NIOSH. 2006c. Health hazard evaluation and technical assistance procedures manual (revised August 2006). Provided to the committee by HHE Program staff.

NIOSH. 2006d. New Orleans Fire Department, New Orleans, Louisana. S. Tak and C. H. Dowell. NIOSH Health Hazard Evaluation Report: HETA 2006-0023-3003. http://www.cdc.gov/niosh/hhe/reports/pdfs/2006-0023-3003.pdf (accessed July 14, 2008).

NIOSH. 2006e. New Orleans Police Department, New Orleans, Louisiana. C. A. West, B. P. Bernard, M. Kitt, C. Mueller, S. Tak, R. J. Driscoll, and J. Hurrell. NIOSH Health Hazard Evaluation Report: HETA 2006-0027-3001. http://www.cdc.gov/niosh/hhe/reports/pdfs/2006-0027-3001.pdf (accessed March 28, 2008).

NIOSH. 2006f. NIOSH letter report: HETA 2006-0120 (dated April 24). S. E. Brueck. Provided in NIOSH response #1 to questions for the NIOSH HHE Program from the Committee to Evaluate the NIOSH Health Hazard Evaluation Program. Fifth example for question #23. Cincinnati, OH.

NIOSH. 2006g. Sara Lee Food, Storm Lake, Iowa. B. King, E. Page, and A. Khan. NIOSH Health Hazard Evaluation Report: HETA 2006-0153-3022. http://www.cdc.gov/niosh/hhe/reports/pdfs/2006-0153-3022.pdf (accessed July 25, 2008).

NIOSH. 2006h. Silicosis—Working with cement roofing tiles: A silica hazard. NIOSH Publication No. 2006-110. http://www.cdc.gov/niosh/docs/2006-110/ (accessed July 25, 2008).

NIOSH. 2006i. Transportation Security Administration, Cincinnati/Northern Kentucky International Airport, Erlanger, Kentucky. C. H. Dowell and A. M. Markey. NIOSH Health Hazard Evaluation Report: HETA 2005-0197-3010. http://www.cdc.gov/niosh/hhe/reports/pdfs/2005-0197-3010.pdf (accessed July 25, 2008).

NIOSH. 2006j. University of Dayton Research Institute, Dayton, Ohio. M. M. Methner, M. E. Birch, D. Evans, and M. D. Hoover. NIOSH Health Hazard Evaluation Report: HETA 2005-0291-3025. http://www.cdc.gov/niosh/hhe/reports/pdfs/2005-0291-3025.pdf (accessed March 28, 2008).

NIOSH. 2007a. Close-out of HETA 2007-0117. Memorandum to HETA File 2007-0117, from team lead industrial hygienist, HETAB, SDHEFS (R-11), through B. Bernard and T. A. Seitz, October 9.

NIOSH. 2007b. NIOSH Health Hazard Evaluation Program: Evidence package for National Academies' review. Produced by the HHE Program staff for the Committee to Review the NIOSH Health Hazard Evaluation Program. Cincinnati, OH.

NIOSH. 2007c. HHE Program staff presentations to the Committee to Evaluate the NIOSH Health Hazard Evaluation Program, October 18, Washington, DC.

NIOSH. 2007d. NIOSH letter report: HETA 2007-0318 (dated September 14). M. Martin. Provided in NIOSH response #1 to questions for the NIOSH HHE Program from the Committee to Evaluate the NIOSH Health Hazard Evaluation Program. First example for question #23. Cincinnati, OH.

NIOSH. 2007e (November 14). NIOSH response #1 to questions for the NIOSH HHE Program from the Committee to Evaluate the NIOSH Health Hazard Evaluation Program. Cincinnati, OH.

NIOSH. 2007f (November 27). NIOSH response #2 to questions for the NIOSH HHE Program from the Committee to Evaluate the NIOSH Health Hazard Evaluation Program. Cincinnati, OH.

NIOSH. 2007g (December 19 [revised]). NIOSH response #3 to questions for the NIOSH HHE Program from the Committee to Evaluate the NIOSH Health Hazard Evaluation Program. Cincinnati, OH.

NIOSH. 2007h (December 21). NIOSH response #4 to questions for the NIOSH HHE Program from the Committee to Evaluate the NIOSH Health Hazard Evaluation Program. Cincinnati, OH.

NIOSH. 2007i. Report on severe fixed obstructive lung disease in workers at a flavoring manufacturing plant. Carmi Flavor and Fragrance Company, Inc., Commerce, California. R. Kanwal and G. Kullman. Health Hazard Evaluation Report: HETA 2006-0303-3043. http://www.cdc.gov/niosh/hhe/reports/pdfs/2006-0303-3043.pdf (accessed March 21, 2008).

NIOSH. 2007j. Simple solutions for construction workers. J. T. Albers and C. F. Estill. NIOSH Publication No. 2007-122. http://www.cdc.gov/niosh/docs/2007-122/pdfs/2007-122-full.pdf (accessed March 28, 2008).

NIOSH. 2008a. Conference call with HHE Program staff R. Sinclair, T. Schnorr, and A. Tepper; committee members R. Henderson and J. Cone; and National Academies staff. February 5.

NIOSH. 2008b. HHE Program staff oral communication with the Committee to Evaluate the NIOSH Health Hazard Evaluation Program, January 15, Washington, DC.

NIOSH. 2008c (January 15 [revised]). NIOSH HHE decision process (triage flow chart) submitted in response to questions for the NIOSH HHE Program from the Committee to Evaluate the NIOSH Health Hazard Evaluation Program during the third committee meeting. Cincinnati, OH.

NIOSH. 2008d (January 30). NIOSH response #6 to questions for the NIOSH HHE Program from the Committee to Evaluate the NIOSH Health Hazard Evaluation Program. Cincinnati, OH.

NIOSH. 2008e (June 3). NIOSH response #7 to questions for the NIOSH HHE Program from the Committee to Evaluate the NIOSH Health Hazard Evaluation Program. Cincinnati, OH.

NIOSH BSC (Board of Scientific Counselors). 1997 (January 2). HHE Review Team: Revised draft report. Appendix 2.5 in NIOSH Health Hazard Evaluation Program: Evidence package for National Academies' review (NIOSH 2007b, above).

NIOSH BSC (Board of Scientific Counselors). 1997 (January 2). HHE Review Team: Revised draft report. Appendix 2.6 in NIOSH Health Hazard Evaluation Program: Evidence package for National Academies' review (NIOSH 2007b, above).

NIOSH BSC. 2006 (March 30). HHE Review Team draft report. Appendix 2.6 in NIOSH Health Hazard Evaluation Program: Evidence package for National Academies' review (NIOSH 2007b, above).

NIOSH eNews. 2008. National Academies seeks input for its evaluation of the NIOSH Health Hazard Evaluation Program. http://www.cdc.gov/niosh/enews/enewsV5N10.html#c (accessed March 21, 2008).

NIOSH and OERP (Occupational Energy Research Program). 2001. NIOSH Energy-Related Health Research Program: Previous DOE studies. http://www.cdc.gov/niosh/2001-133h.html (accessed July 26, 2008).

NRC (National Research Council) and IOM (Institute of Medicine). 2001. *Musculoskeletal disorders and the workplace: Low back and upper extremities.* Panel on Musculoskeletal Disorders and the Workplace. Commission on Behavioral and Social Sciences and Education, Washington, DC: National Academy Press.

NRC and IOM. 2007. *Mining safety and health research at NIOSH: Reviews of research programs at the National Institute for Occupational Safety and Health.* Committee to Review the NIOSH Mining Safety and Health Research Program, Committee on Earth Resources. Washington, DC: The National Academies Press.

NTP (National Toxicology Program). 2002a. NTP-CERHR Expert Panel report on the reproductive and developmental toxicity of 1-bromopropane. Center for the Evaluation of Risks to Human Reproduction. Report NTP-CERHR-1-BP-02. http://cerhr.niehs.nih.gov/chemicals/bromopropanes/1-bromopropane/1-BP_final.pdf (accessed July 25, 2008).

NTP. 2002b. NTP-CERHR Expert Panel report on the reproductive and developmental toxicity of 2-bromopropane. Center for the Evaluation of Risks to Human Reproduction. Report NTP-CERHR-2-BP-02. http://cerhr.niehs.nih.gov/chemicals/bromopropanes/2-bromopropane/2-BP_final.pdf (accessed July 25, 2008).

NTP. 2008. About the NTP. http://ntp-server.niehs.nih.gov/?objectid=7201637B-BDB7-CEBA-F57E39896A08F1BB/ (accessed July 25, 2008).

Olíva, J. 2007. Oral discussion with the Committee to Evaluate the NIOSH Health Hazard Evaluation Program, December 10, Washington, DC.

OSHA (Occupational Safety and Health Administration). 1998 (October 6). Hazard information bulletin—Dust explosion hazard in certain textile processes. http://www.osha.gov/dts/hib/hib_data/hib19981006.html (accessed August 10, 2008).

OSHA. 2000 (September 26). Inspection Report No. 303206387. http://defendingscience.org/case_studies/upload/OSHA-Inspection-Report-2000.pdf (accessed July 7, 2008).

OSHA. 2003 (September 9). Anthrax e-tool: Protecting the worksite against terrorism. Available in PDF form for printing and off-line use. http://www.osha.gov/SLTC/etools/anthrax/index.html and http://www.osha.gov/SLTC/etools/anthrax/pdf/anthraxtask.pdf (accessed July 15, 2008).

OSHA. 2007 (November 6). Quinone (*p*-benzoquinone) 106-51-4. Technology Transfer Network Air Toxics Website. http://www.epa.gov/ttn/atw/hlthef/quinone.html (accessed July 26, 2008).

Pechter, E. 2008 (February 6). Personal response to questionnaire from the Committee to Review the NIOSH Health Hazard Evaluation Program.

Roberts, D. 2000 (May 19). Letter to M. Gaines, OSHA, from D. Roberts, Director, Section for Environmental and Public Health, Missouri Department of Health. http://defendingscience.org/case_studies/upload/Roberts-Letter.pdf (accessed January 14, 2009).

Robinson, J. C. 1991. Pp. 150-151 in *Toil and toxics.* Berkeley: University of California Press.

Rose, C. 2007 (July 18). Letter from C. Rose, M.D., M.P.H., to Mr. M. Landa, Deputy Commissioner for Regulatory Affairs, Office of the Center Director, Center for Food Safety and Applied Nutrition, Food and Drug Administration, College Park, MD, regarding identification of a patient with inhalational exposure as a heavy, daily consumer of butter flavored microwave popcorn. National Jewish Medical and Research Center, Denver, CO. http://www.fda.gov/ohrms/dockets/dockets/06p0379/06p-0379-let0002-01-vol2.pdf (accessed September 8, 2008).

Rosenstock, L. 1996. NIOSH testimony to the U.S. Department of Labor on indoor air quality. *Applied Occupational and Environmental Hygiene 11*(12):1365-1370.

RTI (Research Triangle Institute). 1997 (December). Development of an evaluation system to assess NIOSH's Health Hazard Evaluation Program—Final report. J. Hersey, A. Boshes, L. H. Kojetin, T. Smith, T. Wilcosky, and B. B. Burrus. Technical Report. Contract No. CDC-200-93-0697. Washington, DC: RTI.

Sandia National Laboratories. 2005 (April 21). Sandia BROOM toll can help restore facilities following release of biological warfare agents. News Release. http://www.sandia.gov/news-center/news-releases/2005/def-nonprolif-sec/broom.html (accessed July 15, 2008).

Schneider, A. 2007 (September 5). Popcorn supplier to drop toxic chemical: Consumer's lung disease may be linked to flavoring. *Seattlepi.com* (online). http://seattlepi.nwsource.com/food/330230_popcorn05.html?source=mypi (accessed October 10, 2008).

Schneider, A. 2008 (January 4). Flavoring additive puts professional cooks at risk. *Seattle Post-Intelligencer* (online). http://seattlepi.nwsource.com/national/344277_diacetyl21.html (accessed April 16, 2008).

Sharfstein, J. 2007. Oral discussion with the Committee to Evaluate the NIOSH Health Hazard Evaluation Program, December 10, Washington, DC.

Sheehan, M. J. 1999 (July 15). Summary: Final report of the OSHA Metalworking Fluids Standard Advisory Committee. http://www.osha.gov/dhs/reports/metalworking/MWFSAC-FinalReport Summary.pdf (accessed April 2, 2008).

Silverstein, B., and D. Adams. 2007. Work-related musculoskeletal disorders in the neck, back, and upper extremity in Washington State, 1997-2005. Technical Report Number 40-11. Olympia, WA: Safety & Health Assessment & Research for Prevention (SHARP) and Washington State Department of Labor and Industries.

Speakman, R. 2008. Oral discussion with the Committee to Evaluate the NIOSH Health Hazard Evaluation Program, January 15, Irvine, CA.

Spencer, P. S., and H. H. Schaumburg. 2000. *Experimental and clinical neurotoxicology, 2nd edition.* New York: Oxford University Press.

Stakeholder Response Table. 2008. Responses to questionnaire distributed electronically by and posted on a website created for the Committee to Evaluate the NIOSH Health Hazard Evaluation Program. Review of the NIOSH Health Hazards Evaluation Program, DELS-O-04-01-G. Available from the Public Access Records Office (PARO@nas.edu) of the National Academies.

Steenhuysen, J. 2007 (September 5). FDA to probe popcorn link in man's lung disease. *Reuters UK* (online). http://uk.reuters.com/article/healthNews/idUKN0524254120070905 (accessed July 23, 2008).

Trout, D., A. Nimgade, C. Mueller, R. Hall, and G. S. Earnest. 2002. Health effects and occupational exposures among office workers near the World Trade Center disaster site. *Journal of Occupational and Environmental Medicine* 44(7):601-605.

U.S. Coast Guard. 2001 (February 28). Media Advisory: U.S.C.G. recalls houseboats due to carbon monoxide hazard and cautions owners/operators on dangers. USCG Headquarters Public Affairs, Press Release. Washington, DC. http://www.uscgboating.org/articles/boatingview.aspx?id=18 (accessed July 28, 2008).

U.S. Coast Guard. 2008. Carbon monoxide advisory: Generator exhaust. Washington, DC. http://www.uscgboating.org/alerts/alertsview.aspx?id=8 (accessed July 28, 2008).

U.S. Congress, House of Representatives, Committee on Appropriations. 2007. Agriculture, Rural Development, Food and Drug Administration, and Related Agencies Appropriations Bill, 2008 (H.R. 110-258). 110th Cong., 1st Sess., July 24. http://frwebgate.access.gpo.gov/cgibin/getdoc.cgi?dbname= 110_cong_reports&docid=f:hr258.110.pdf (full text of bill) (accessed July 23, 2008).

U.S. Congress, House of Representatives, House Education and Labor Committee. 2007. Popcorn Workers Lung Disease Prevention Act (H.R. 2693 PCS). 110th Cong., 1st Sess., June 13.

U.S. Congress, Senate, Committee on Labor and Public Welfare. 1968. Statement of P. R. Lee, M.D., Assistant Secretary, Health and Scientific Affairs, Department of Health, Education, and Welfare, 90th Cong., 2nd Sess., June 19. Pp. 269-273 in published hearing on the Occupational Safety and Health Act of 1968, S. 2864, February 15; June 12, 19, 24, 28; and July 2. Washington, DC: U.S. Government Printing Office.

U.S. Department of Homeland Security. 2008. National Response Framework, January 2008. http://www.fema.gov/pdf/emergency/nrf/nrf-core.pdf (accessed July 8, 2008). *Note:* This core document, along with the Emergency Support Function Annexes and Support Annexes (available at the NRF Resource Center, http://www.fema.gov/NRF), supersedes the corresponding sections of the *National Response Plan* (2004, with 2006 revisions).

U.S. Department of Labor. 2000. Ergonomics Program; Final Rule. 65:68261-68870. 20 CFR Part 1910. Occupational Safety and Health Administration. *Federal Register* (archived). http://www.osha.gov/FedReg_osha_pdf/FED20001114.pdf (accessed August 10, 2008).

U.S. Department of Labor. 2008a. Regulations (Standards–29 CFR). Table of contents–1910. Occupational Safety and Health Administration, www.osha.gov. http://www.osha.gov/pls/oshaweb/owadisp.show_document?p_table=STANDARDS&p_id=9696 (accessed July 25, 2008).

U.S. Department of Labor. 2008b. Regulations (Standards–29 CFR). Table of contents–Lead–1926.62. http://www.osha.gov/pls/oshaweb/owadisp.show_document?p_table=STANDARDS&p_id=10641 (accessed July 25, 2008).

U.S. Department of Labor. 2008c. Title 30, Code of federal regulations. 30 CFR § 71.700. Inhalation hazards; threshold limit values for gases, dust, fumes, mists, and vapors. Mine Safety and Health Administration, www.msha.gov. http://www.msha.gov/30cfr/71.700.htm (accessed July 25, 2008).

U.S. National Park Service. 2005 (July 26). Advisory: Carbon monoxide poisoning hazards cause by boating operations. Washington Office Concession Program & Division of Risk Management, Washington, DC. http://www.concessions.nps.gov/images/webpages/CarbonMonoxide Advisory07.26.05-CPC-FINAL.pdf (accessed July 28, 2008).

Washington State Department of Health. 2008. Occupational Mortality Database. http://www3.doh.wa.gov/OCCMORT/ (accessed March 26, 2008).

Washington State Department of Labor & Industries. 2000. Concise Explanatory Statement (RCW 34.05.325.6a), WAC 296-62-051, Ergonomics. http://www.lni.wa.gov/wisha/ergo/rule_docs/ces/CES.524.PDF (accessed July 23, 2008).

Washington State Department of Labor & Industries. 2008a (February). Do you cook with butter-flavored oils or butter substitutes? Information for restaurant owners and workers about diacetyl. http://www.lni.wa.gov/Safety/REsearch/Files/DiacetylFactSheet.pdf (accessed July 23, 2008).

Washington State Department of Labor & Industries. 2008b. OSHA, California and Washington rule making records. http://www.lni.wa.gov/Safety/Topics/Ergonomics/History/Documents/default.asp (accessed March 28, 2008).

Worksafe. 2007. Occupational Safety and Health Conference: Developing a Pro-Worker OSH Public Policy Agenda: Learning from the Past—Building for the Future, December 2-3, Emeryville, California. http://urbanhabitat.org/sec/newsletter/dec2007/events/01-worksafe/ (accessed April 2, 2008).

Appendixes

A

Framework for the Review of Research Programs of the National Institute for Occupational Safety and Health*

This is the second version of a document prepared by the National Academies Committee for the Review of NIOSH Research Programs[1] also referred to as the Framework Committee. This document is not a formal report of the National Academies—rather, it is a framework proposed for use by multiple National Academies evaluation committees to review up to 15 National Institute for Occupational Safety and Health (NIOSH) research programs. It is a working document subject to modification by the Framework Committee on the basis of responses received from evaluation-committee members, NIOSH, stakeholders, and the general public during the course of the assessments.

---

*Version of 8/10/07.

[1]Members of the committee at the time this version was produced were David Wegman, *Chair* (University of Massachusetts Lowell School of Health and Environment), William Bunn III (International Truck and Engine Corporation), Carlos Camargo (Harvard Medical School), Susan Cozzens (Georgia Institute of Technology), Letitia Davis (Massachusetts Department of Public Health), James Dearing (Kaiser Permanente-Colorado), Fred Mettler, Jr. (University of New Mexico School of Medicine), Franklin Mirer (Hunter College School of Health Sciences), Jacqueline Nowell (United Food and Commercial Workers International Union), Raja Ramani (Pennsylvania State University), Jorma Rantanen (International Commission on Occupational Health), Rosemary Sokas (University of Illinois at Chicago School of Public Health), Richard Tucker (Tucker and Tucker Consultants, Inc., and University of Texas at Austin), and James Zuiches (North Carolina State University). Sammantha Magsino (National Academies staff) was the study director. Joseph Wholey (University of Southern California), former committee member, contributed to the first version of this document. Part V includes brief biographies of current committee members.

This version reflects several significant changes to the original framework document (version 12/19/05) that was used to guide the work of the first four evaluation committees (Hearing Loss; Mining; Agriculture, Forestry, and Fishing; and Respiratory Disease). Changes were made in response to feedback from members and staff of these committees, as well as other comments on the original framework, in order to make the document more useful to evaluation committees as they carry out their work. In particular, the following changes were made to the framework document during the revision process:

- The wording of some of the relevance and impact scores were edited to make the wording more precise and to reduce situations where the original scores were non-unique or overlapping (revised scoring criteria are given in Boxes 2 and 3).
- A new table was added to provide explicit guidance to evaluation committees on how to weigh differences in the observed levels of "research priority" and "engagement in appropriate transfer activities" in arriving at a single integer score for relevance (see Table 6).
- The guidance on scoring was clarified to make more explicit that all scores are to be given as integers.
- The NIOSH logic model was updated (see Figure 1).
- The table on evaluation committee information needs (Table 2) was reorganized to be more consistent with the NIOSH logic model, and additional information needs identified by the first set of evaluation committees were added.
- A worksheet to assist with the development of scores has been deleted and key components of the worksheet have been incorporated into appropriate sections throughout the document.
- The organization of the document was modified to more closely follow the revised statement of task and to improve readability.
- A number of sections of text originally presented in outline form were modified in tables or boxes to make the information more accessible.

This second version of the framework document remains a working document subject to further modification by the Framework Committee on the basis of input received from evaluation committee members, NIOSH, stakeholders, and the general public during the course of the assessments.

CONTENTS

ABBREVIATIONS AND ACRONYMS

ABLES	Adult Blood Lead Epidemiology and Surveillance
AOEC	Association of Occupational and Environmental Clinics
BLS	Bureau of Labor Statistics
CDC	Centers for Disease Control and Prevention
CSTE	Council of State and Territorial Epidemiologists
DOD	U.S. Department of Defense
EC	Evaluation Committee
EPA	Environmental Protection Agency
FACE	Fatality Assessment Control and Evaluation
FC	Framework Committee
HHE	Health Hazard Evaluation
MSHA	Mine Safety and Health Administration
NIH	National Institutes of Health
NIOSH	National Institute for Occupational Safety and Health
NORA	National Occupational Research Agenda
NORA1	National Occupational Research Agenda 1996-2005
NORA2	National Occupational Research Agenda 2005-forward
OSH Review Commission	Occupational Safety and Health Review Commission
OSHA	Occupational Safety and Health Administration
OSHAct	Occupational Safety and Health Act of 1970
PART	Performance Assessment Rating Tool
PEL	permissible exposure limit
RFA	request for applications
SENSOR	Sentinel Event Notification System of Occupational Risks

TMT tools, methods, or technologies

USDA U.S. Department of Agriculture

I. INTRODUCTION

In September 2004, the National Institute for Occupational Safety and Health (NIOSH) contracted with the National Academies to conduct a review of NIOSH research programs. The goal of this multiphase effort is to assist NIOSH in increasing the impact of its research efforts that are aimed at reducing workplace illnesses and injuries and improving occupational safety and health. The National Academies assigned the task to the Division on Earth and Life Studies and the Institute of Medicine.

The National Academies appointed a committee of 14 members, including persons with expertise in occupational medicine and health, industrial health and safety, industrial hygiene, epidemiology, civil and mining engineering, sociology, program evaluation, communication, and toxicology; representatives of industry and of the workforce; and a scientist experienced in international occupational-health issues. The Committee on the Review of NIOSH Research Programs, referred to as the Framework Committee (FC), prepared the first version of this document during meetings held on May 5-6, July 7-8, and August 15-16, 2005. This second version was finalized after the Framework Committee's May 30-31, 2007, meeting, based on feedback received on the framework from the first two independent evaluation committees, NIOSH leadership, and National Academies' staff, as well as discussions during an earlier FC meeting in April 2006.

This document is not a report of the National Academies; rather, it presents the evaluation framework developed by the FC to guide and provide common structure for the reviews of as many as 15 NIOSH programs during a 5-year period by independent evaluation committees (ECs) appointed by various divisions and boards of the National Academies. It is a working document to be shared with NIOSH and the public. This version has been modified by the FC on the basis of responses from the ECs, NIOSH, NIOSH stakeholders, and the public; and it may be modified again. It is incumbent on the ECs to consult with the FC if portions of the evaluation framework presented here are inappropriate for specific programs under review.

I.A. Overview of Charge to Evaluation Committees

At the first meeting of the FC, Lewis Wade, NIOSH senior science adviser, emphasized that a review of a NIOSH program should focus on the program's relevance to and impact on health and safety in the workplace. In developing a framework, the FC considered the following elements of the charge to the ECs:

1. Assessment of the program's contribution, through occupational safety and health research, to reductions in workplace hazardous exposures, illnesses, or injuries through

   a. an assessment of the relevance of the program's activities to the improvement of occupational safety and health, and
   b. an evaluation of the impact that the program's research has had in reducing work-related hazardous exposures, illnesses, and injuries.

   The evaluation committee will rate the performance of the program for its relevance and impact using an integer score of 1-5. Impact may be assessed directly (for example, on the basis of reductions in illnesses or injuries) or, as necessary, by using intermediate outcomes to estimate impact. Qualitative narrative evaluations should be included to explain the numerical ratings.
2. Assessment of the program's effectiveness in targeting new research areas and identifying emerging issues in occupational safety and health most relevant to future improvements in workplace protection. The committee will provide a qualitative narrative assessment of the program's efforts and suggestions about emerging issues that the program should be prepared to address.

I.B. Evaluation Committees

Individual ECs will be formed in accordance with the rules of the National Academies for the formation of balanced committees. Each EC will comprise persons with expertise appropriate for the specific NIOSH research program under review and may include representatives of stakeholder groups (such as labor unions and industry), experts in technology and knowledge transfer, and program evaluation. The EC will gather appropriate information from the sponsor (the NIOSH research program under review), stakeholders affected directly by NIOSH program research, and relevant independent parties. Each EC will consist of about 10 members, will meet about three times, and will prepare a report. The National Academies will deliver the report to NIOSH within 9 months of the first meeting of the EC. EC reports are subject to the National Academies report-review process.

I.C. NIOSH Strategic Goals and Operational Plan

As a prelude to understanding the NIOSH strategic goals and operational plan, NIOSH research efforts should be understood in the context of the Occupational Safety and Health Act (OSHAct), under which it was created. The OSHAct identifies workplace safety and health as having high national priority and gives employers the responsibility for controlling hazards and preventing workplace injury and illness. The act creates an organizational framework for doing that, assigning complementary roles and responsibilities to employers and employees, the Occupational Safety and Health Administration (OSHA), the states, the Occupational Safety and Health (OSH) Review Commission, and NIOSH. The act recognizes NIOSH's role and responsibilities to be supportive and indirect. NIOSH research, training programs, criteria, and recommendations are intended to be used to inform and assist those more directly responsible for hazard control (OSHAct Sections 2b, 20, and 22).

Section 2b of the OSHAct describes 13 interdependent means of accomplishing the national goal, one of which is "by providing for research . . . and by developing innovative methods . . . for dealing with occupational safety and health problems." Sections 20 and 22 give the responsibility for that research to NIOSH. NIOSH is also given related responsibilities, including the development of criteria to guide prevention of work-related injury or illness; development of regulations for reporting on employee exposures to harmful agents; establishment of medical examinations, programs, or tests to determine illness incidence and susceptibility; publication of a list of all known toxic substances; assessment of potential toxic effects or risks associated with workplace exposure in specific settings; and conduct of education programs for relevant professionals to carry out the OSHAct purposes. NIOSH is also responsible for assisting the secretary of labor regarding education programs for employees and employers in hazard recognition and control.

The NIOSH mission is "to provide national and world leadership to prevent work-related illness, injury, disability, and death by gathering information, conducting scientific research, and translating the knowledge gained into products and services." To fulfill its mission, NIOSH has established the following strategic goals:[2]

- **Goal 1: Conduct research to reduce work-related illnesses and injuries.**
    - Track work-related hazards, exposures, illnesses, and injuries for prevention.
    - Generate new knowledge through intramural and extramural research programs.

[2]See http://www.cdc.gov/niosh/docs/strategic/.

    - Develop innovative solutions for difficult-to-solve problems in high-risk industrial sectors.

- **Goal 2: Promote safe and healthy workplaces through interventions, recommendations, and capacity building.**
    - Enhance the relevance and utility of recommendations and guidance.
    - Transfer research findings, technologies, and information into practice.
    - Build capacity to address traditional and emerging hazards.

- **Goal 3: Enhance global workplace safety and health through international collaborations.**
    - Take a leadership role in developing a global network of occupational health centers.
    - Investigate alternative approaches to workplace illness and injury reduction and provide technical assistance to put solutions in place.
    - Build global professional capacity to address workplace hazards through training, information sharing, and research experience.

In 1994, NIOSH embarked on a national partnership effort to identify research priorities to guide occupational health and safety research for the next decade. The National Occupational Research Agenda (NORA) identified 21 high-priority research subjects (see Table 1). The NORA was intended not only for NIOSH but for the entire occupational health community. In the second decade of the NORA, NIOSH is working with its partners to update the research agenda, using an approach based on industry sectors. NIOSH and its partners are working through sector research councils to establish sector-specific research goals and objectives. The emphasis is on moving research to practice in workplaces through sector-based partnerships.

Figure 1 is the NIOSH operational plan, presented as a logic model,[3] of the path from inputs to outcomes for each NIOSH research program. The FC adapted the model to develop its framework. NIOSH will provide similar logic models appropriate to each research program evaluated by an EC.

[3]Developed by NIOSH with the assistance of the RAND Corporation.

TABLE 1 NORA High-Priority Research by Category

Category	Priority Research Area
Disease and injury	Allergic and irritant dermatitis
	Asthma and chronic obstructive pulmonary disease
	Fertility and pregnancy abnormalities
	Hearing loss
	Infectious diseases
	Low-back disorders
	Musculoskeletal disorders of upper extremities
	Trauma
Work environment and workforce	Emerging technologies
	Indoor environment
	Mixed exposures
	Organization of work
	Special populations at risk
Research tools and approaches	Cancer research methods
	Control technology and personal protective equipment
	Exposure-assessment methods
	Health-services research
	Intervention-effectiveness research
	Risk-assessment methods
	Social and economic consequences of workplace illness and injury
	Surveillance research methods

I.D. Evaluation Committees' Information Needs

Each NIOSH program under review will provide information to the relevant EC, including that outlined in Table 2. The EC may request additional information of NIOSH as needed, and NIOSH should provide it as quickly as is practical. NIOSH should consider organizing the information listed in Table 2 by subprogram or program as appropriate and to the extent possible.

In addition to the information provided by NIOSH, the EC should independently collect additional information that it deems necessary for evaluation (for example, the perspectives of stakeholders, such as OSHA, MSHA, unions and workforces, and industry). In conducting the review, the EC should continually examine how individual projects or activities contribute to the impact and relevance of a program as a whole.

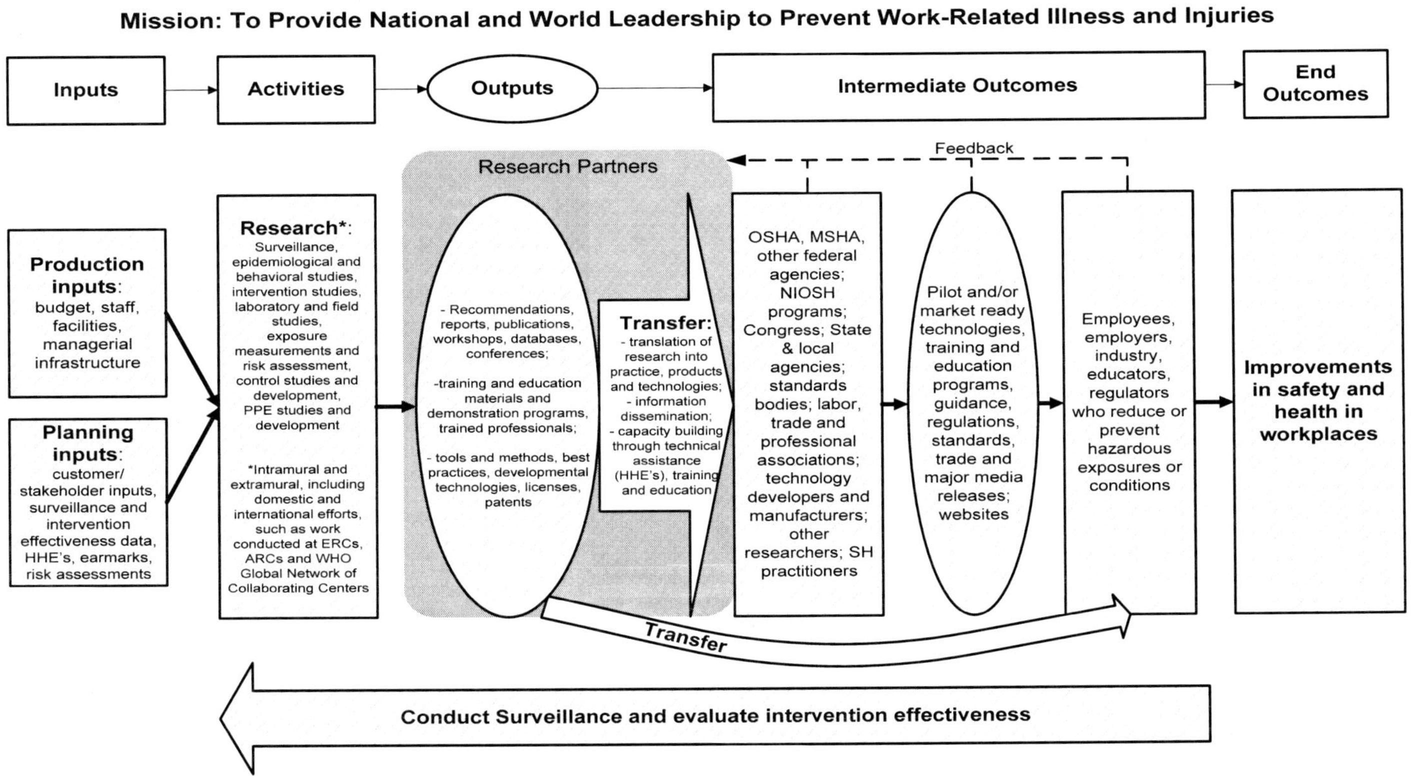

FIGURE 1 The NIOSH operational plan presented as a logic model.

TABLE 2 Evaluation Committee Information Needs

- **Program background and resources:**
  - Program history.
  - Major program challenges.
  - Program strategic goals and objectives, past (for period under review) and current.
  - Major subprograms (if appropriate).
  - Results of previous program reviews (for example, annual review by NIOSH leadership team or external scientific program reviews).
  - External factors affecting the program.

- **Interactions with stakeholders and with other NIOSH programs:**
  - The role of program research staff in NIOSH policy-setting, OSHA and MSHA standard-setting, voluntary standard-setting and other government policy functions.
  - Interactions and working relationships with other NIOSH programs.
  - Identification of other institutions and research programs with overlapping or similar portfolios and an explanation of the relationship between NIOSH activities and those of other institutions.
  - Key partnerships with employers, labor, other government organizations, academic institutions, nonprofit organizations, and international organizations.

- **Program inputs:**
  - Program resources (also called *production inputs*).
    - Funding by year for period under review.
    - Funding by objective or subprogram.
    - Program staffing, FTE's, and laboratory facilities, by subprogram (if indicated).
    - Percentage of program budget that is discretionary (beyond salaries).
    - Percentage of program budget that is earmarked.
    - Contributions from other agencies (in kind or funds).
  - Planning inputs.
    - Surveillance data, inputs from the Health Hazard Evaluation (HHE) or Fatality Assessment Control and Evaluation (FACE) program, or intramural and extramural research findings that influenced program goals and objectives.
    - Planning inputs from stakeholders, for example, advisory groups, NORA teams, and professional, industry, and labor groups (specify if any input from groups representing small business or vulnerable populations).
    - Related OSHA, Mine Safety and Health Administration (MSHA) strategic plans, or other input.
    - Process for soliciting and approving intramural research ideas.
    - Process for soliciting and approving program-supported extramural research activities.

- **Program activities (more details provided in Table 3):**
  - Intramural.
    - Surveillance activities.
    - Research activities (projects).

continued

TABLE 2 Continued

- Transfer activities to encourage implementation of research results for improved occupational safety and health (for example, information dissemination, technical assistance, and technology and knowledge transfer).
- Key collaborations in intramural activities (for example, with other government agencies, academe, industry, and unions).

- Extramural funded by NIOSH.
  - Requests for applications (RFAs) developed by program.
  - Funded projects: grants, cooperative agreements, and contracts, such as
    - ◊ Surveillance activities.
    - ◊ Research activities.
    - ◊ Transfer activities.
    - ◊ Capacity-building activities.

- **Outputs (products of the research program—more details provided in Table 4):**
  - Intramural.
    - Peer-reviewed publications, agency reports, alerts, and recommendations.
    - Databases, Web sites, tools, and methods (including education and training materials).
    - Technologies developed and patents.
    - Sponsored conferences and workshops.
  - Extramural (to the extent practical).

- **Intermediate outcomes:**
  - Standards or guidelines issued by other agencies or organizations based in whole or in part on NIOSH research.
  - Adoption and use of control or personal protective technologies developed by NIOSH.
  - Evidence of industry, employer, or worker behavioral changes in response to research outputs.
  - Use of NIOSH products by workers, industry, occupational health and safety professionals, health care providers, and so on (including internationally).
  - NIOSH Web-site hits and document requests.
  - Unique staff or laboratory capabilities that serve as a national resource.
  - Other intermediate outcomes.

- **End outcomes:**
  - Data on program impact on rates and numbers of injuries and illnesses and exposures in the workplace (including trend data, if available).
  - Documentation of workplace risk reduction (quantitative, qualitative, or both).

- **Description of current processes for setting research priorities and identifying emerging issues in the workplace.**

I.E. Prior Evaluations

Several NIOSH programs have already been evaluated by internal and external bodies. The evaluations may have been part of an overall assessment of NIOSH, such as the 2005 Performance Assessment Rating Tool (PART) review,[4] or the evaluation of specific research program elements, such as any external scientific-program review. NIOSH should inform of, and the ECs should review, all prior evaluations of the program under review as an aid to understanding the evolution of the program and its elements. The EC evaluations, however, are independent of prior reviews and evaluations.

II. SUMMARY OF EVALUATION PROCESS

The ECs will assess the relevance and impact of NIOSH research programs. In conducting their evaluations, the ECs should ascertain whether NIOSH is doing the right things (relevance) and whether these things are improving health and safety in the workplace (impact).

II.A. The Evaluation Flowchart (Figure 2)

To address its charge, the FC simplified the logic model of Figure 1 into a flowchart (Figure 2) that breaks the NIOSH logic model into discrete, sequential program components to be assessed by the EC. Each component of Figure 2 is addressed in greater detail in the indicated section of this document. The FC understands that the activities of any research program will not be as linear as presented in either Figures 1 or 2. The major components to be evaluated are

- major program *challenges,*
- strategic *goals and objectives,*
- *inputs* (such as budget, staff, facilities, the institute's research management, the NIOSH Board of Scientific Counselors, the NORA process, and NORA work groups),
- *activities* (efforts by NIOSH staff, contractors, and grantees, such as hazard surveillance; surveillance for injury, illness, and biomarkers of effect; exposure-measurement research; safety-systems research; injury-prevention research; health-effects research; intervention

[4]The PART focuses on assessing program-level performance and is one of the measures of success of the budget and performance integration initiative of the president's management agenda (see CDC Occupational Safety and Health at http://www.whitehouse.gov/omb/budget/fy2006/pma/hhs.pdf).

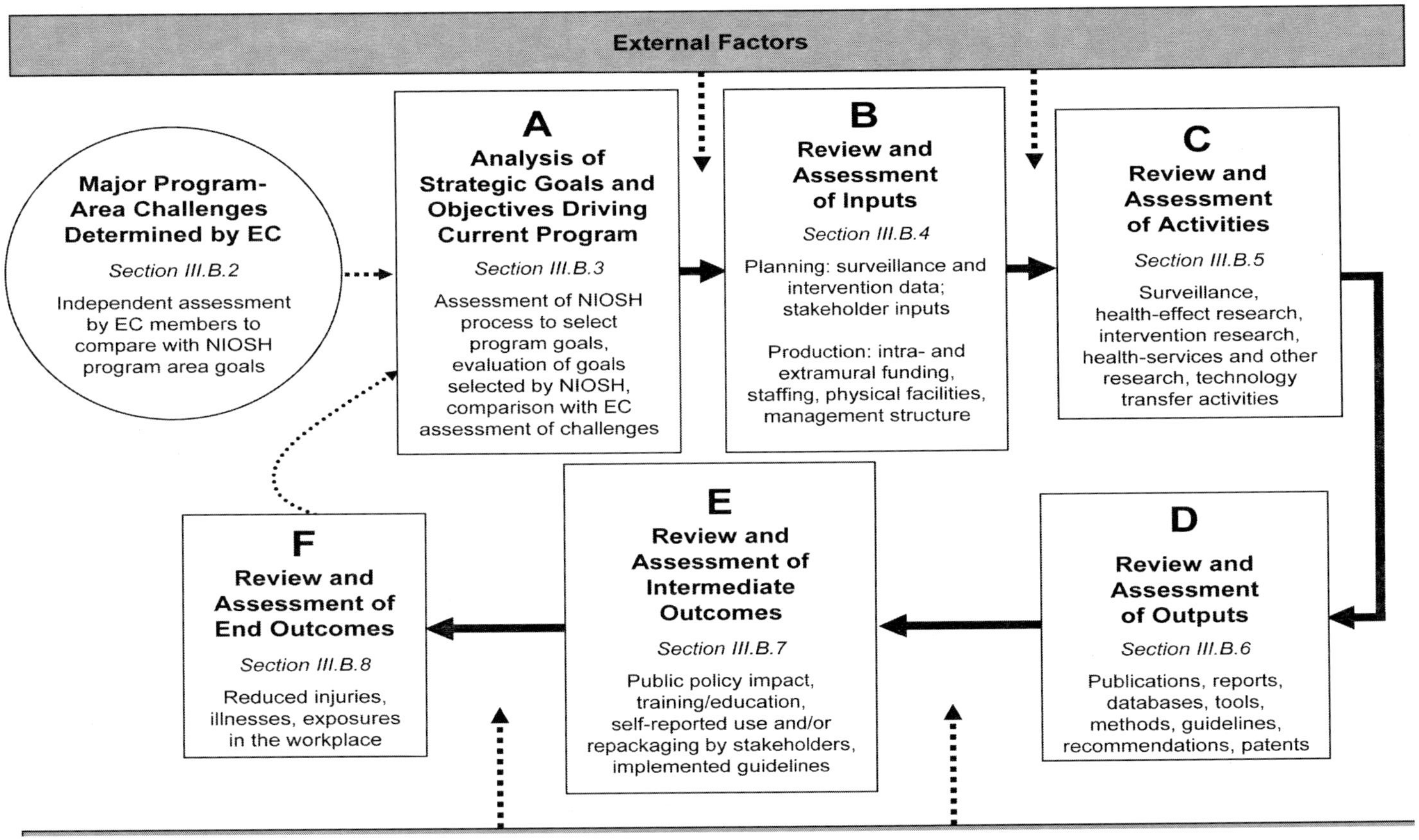

FIGURE 2 Flowchart for the evaluation of the NIOSH research program.

research; health-services research; and technology and knowledge transfer activities),

- *outputs* (NIOSH products, such as publications, reports, conferences, databases, tools, methods, guidelines, recommendations, education and training, and patents),
- *intermediate outcomes* (responses by NIOSH stakeholders to NIOSH products, such as public or private policy change, training and education in the form of workshop or seminar attendance, self-reported use or repackaging of NIOSH data by stakeholders, adoption of NIOSH-developed technologies, implemented guidelines, licenses, and reduction in workplace hazardous exposure), and
- *end outcomes* (such as reduction in work-related injuries or illnesses or hazardous exposures in the workplace).

The flowchart summarizes the FC's vision of how a program evaluation should occur. In evaluating each program or major subprogram, the EC must collect, analyze, and evaluate information on items described in each of the boxes of Figure 2, regardless of management structure (such as linear or matrix). The FC recognizes that the components of any program will not fit perfectly in any category in Figure 1 or 2. For example, training and development programs were appropriately defined as outputs by NIOSH in the logic model (Figure 1), but the FC finds more value in focusing on the responses to these outputs as intermediate outcomes (Figure 2, Box E) in the flowchart. The committee further recognizes that matrix organizations may have little control over the input portion of the logic model and that matrix program management may have fewer resources of its own on which to base its decisions. Following the suggested evaluation procedures, however, should ensure a desired level of consistency and comparability among all the ECs.

Drawing on the program logic model, the flowchart, and EC members' expertise, the ECs will delineate important inputs and external factors affecting the NIOSH research program's agenda and the consequences of NIOSH research activities. Examples of external factors are research activities of industry and other federal agencies and the political and regulatory environment. For purposes of this review, the results of inputs and external factors are the program research activities, outputs, and associated transfer activities that may result in intermediate outcomes and possibly end outcomes.

II.B. Steps in Program Evaluation

The FC concludes that useful evaluation requires a disciplined focus on a small number of questions or hypotheses typically related to program goals, performance criteria, and performance standards; a rigorous method of answering the questions or testing the hypotheses; and a credible procedure for developing qualitative and quantitative assessments. The evaluation process developed by the FC is summarized in Box 1 and described in detail in Section III of this document.

III. EVALUATION OF A NIOSH RESEARCH PROGRAM—THE PROCESS

III.A. Analysis of External Factors Relevant to the NIOSH Research Program

As depicted in the logic model (Figure 1), reduction in injury and illness (end outcomes) or in exposure (intermediate outcome) is affected by stakeholder activities (external factors). Actions of those in labor, industry, regulatory entities, and others beyond NIOSH's control are necessary for the implementation of NIOSH recommendations. Implementation of research findings may depend on existing or future policy considerations.

External factors may be considered as forces beyond the control of NIOSH that may affect the evolution of a program. External factors influence NIOSH's progress through all phases of the logic model and flowchart; from inputs to end outcomes (see Figures 1 and 2). Identification of external factors by an EC is essential because it provides the context for evaluation of the NIOSH program. External factors may be best assessed on the basis of the expert judgment of EC members who have knowledge of the field of research. Information regarding external factors should also be sought from NIOSH, OSHA, and MSHA staff and from other stakeholders. The EC, however, may choose additional approaches to assess external factors. NIOSH should identify and describe external factors early in the evaluation sequence (see Table 2). Factors external to NIOSH might have been responsible for achieving some outcomes or might have presented formidable obstacles. The EC must address both possibilities.

Some external factors may involve constraints on research activities related to target populations, methodologic issues, and resource availability. ECs might examine whether or not the following are true:

- Projects addressing a critical health need are technologically feasible. However, a workforce of appropriate size and with appropriate duration and distribution of exposure for measuring a health effect may not exist; for example, no population of workers has been exposed for 30 years to formaldehyde at the current OSHA permissible exposure limit (PEL), so the related cancer mortality cannot yet be directly assessed.
- Research is inhibited because NIOSH investigators are unable to access an adequate study population. Under current policy, NIOSH must

BOX 1
The Evaluation Process

1. Gather appropriate information from NIOSH and other sources (see Table 2).

2. Determine timeframe to be covered in the evaluation (see III.B.1.).

3. Identify major program area challenges and objectives (see III.B.2.).
All NIOSH research programs are designed to be responsive to present or future workplace safety and health issues. Each research program should have its own objectives. Each EC will provide an independent assessment of the major workplace health and safety problems related to the program under review and determine whether they are consistent with the program's stated goals and objectives.

4. Identify subprograms and major projects in the research program.
Each EC must determine how to disaggregate a program to achieve a manageable and meaningful evaluation of its components, and of the overall program. A program may need to be broken down into several recognizable subprograms or major projects if an effective evaluation is to be organized. It may be advantageous for an EC to disaggregate a program into subprograms that NIOSH identifies.

5. Evaluate the subprogram components sequentially (see III.B.2. through III.B.8.), using the flowchart (Figure 2) as a guide.
This will involve a qualitative assessment of each component of the research program. ECs will use professional judgment to answer questions and follow the guidance provided by the FC.

6. Evaluate the research program's potential outcomes that are not yet appreciated (see III.B.9.).

7. Evaluate the important subprogram outcomes specifically for contributions to improvements in workplace safety and health.
Guidance is provided with specific items for consideration (see III.B.10.).

8. Evaluate and score the overall program for *relevance* (see III.B.10.).
Final program ratings will consist of an integer score and discussion of its rationale.

9. Evaluate and score the overall program for *impact* (see III.B.10.).
Final program ratings will consist of an integer score and discussion of its rationale.

10. Identify success in targeting priority research and emerging issues (see III.C.).
The EC should briefly discuss its assessment of the NIOSH program's process for determining priorities for research and emerging workplace issues. The ECs should also independently identify emerging workplace issues for which the NIOSH program under review should be prepared.

11. Prepare report by using the template provided in Section IV as a guide.

either obtain an invitation by management to study a workplace or seek a judicial order to provide authority to enter a worksite. (Co-operation under court order may well be insufficient for effective research.)

- Research is inhibited because the work environment, materials, and historical records cannot be accessed even with management and workforce cooperation.
- Adequate or established methods do not exist for assessing the environment.
- The NIOSH contribution to a particular field of research is reduced because other institutions are working in the same field.
- NIOSH resources are inadequate to tackle key questions.

Evaluation of the impact of NIOSH research outputs on worker health and safety may require consideration of external factors that might impede or aide implementation, measurement, and so on. ECs might consider whether or not the following are true:

- Regulatory end points are unachievable because of obstacles to regulation or because of differing priorities of the regulatory agencies. For example, there may be no implementation of recommendations for improved respiratory protection programs for health-care workers because of enforcement policies or lack of acceptance by the health-care institution administrators.
- A feasible control for a known risk factor or exposure is unimplemented because the costs of implementation are too high or because current economic incentives do not favor such actions.
- End outcomes are unobservable because baseline and continuing surveillance data are not available. For example, the current incidence of occupational noise-induced hearing loss is not known although surveillance for a substantial threshold shift is feasible. (NIOSH conducts surveillance of work-related illnesses, injuries, and hazards, but comprehensive surveillance is not possible with existing resources.)
- Reductions in adverse effects of chronic exposure cannot be measured. For example, 90% of identified work-related mortality is from diseases, such as cancer, that arise only after decades of latency after first exposure; therefore, effects of reducing exposure to a carcinogen cannot be observed in the timeframe of most interventions.
- A promulgated regulation requires a technology that was developed but not widely used.

- Reductions in fatal traumatic injuries occur because more-hazardous manufacturing jobs are replaced by less-hazardous knowledge-based jobs.

III.B. Evaluating NIOSH Research Programs by Using the Flowchart

The FC used the NIOSH logic model (Figure 1) to define the scope and stages of an EC evaluation. The evaluation of the elements in the flowchart (Figure 2) summarizes the FC's vision of how a program evaluation should proceed. FC members also identified numerous possible factors to consider in assessing the relevance of NIOSH research-program components, including the following:

- The severity or frequency of health and safety hazards addressed and the number of people at risk (magnitude) for these hazards.
- The extent to which NIOSH research programs identify and address gender-related issues and issues of vulnerable populations: Vulnerable populations are defined as groups of workers who have biologic, social, or economic characteristics that place them at increased risk for work-related conditions or on whom inadequate data have been collected. Vulnerable populations include disadvantaged minorities, disabled persons, low-wage workers, and non-English-speakers for whom language or other barriers present health or safety risks.
- The extent to which NIOSH research programs address the health and safety needs of small businesses.
- The "life stage" of problems being addressed: As the health effects are understood, efforts should shift to intervention research, from efficacy to intervention, and to intervention-effectiveness research. Gaps in the spectrum of prevention need to be addressed; for example, research on exposure assessment may be necessary before the next intervention steps can be taken.
- The structure, in addition to the content, of the research program: A relevant research program is more than a set of unrelated research projects; it is an integrated program involving interrelated surveillance, research, and transfer activities.
- Appropriate NIOSH consideration of stakeholder input.

The ECs may consider those and other important factors that bear on relevance as they progress through each stage of an evaluation.

The following subsections are intended to guide the EC through the evaluation process and flowchart in Figure 2. Each begins with a definition of the component being evaluated, provides questions for the EC to consider during the course of its evaluation, and provides some guidance regarding the assessment of the component. The FC admittedly provides little guidance regarding the evaluation of programs that are organized in a matrix structure or programs that have large extramural research components. Because of the uniqueness of each program, each EC must determine the most reasonable way to apply the criteria established in this document.

III.B.1. Identifying the Period for Evaluation

By studying materials presented by the NIOSH research program and other sources, the EC will become familiar with the history of the research program being evaluated and its major subprograms, goals, objectives, resources, and other pertinent information. Having that information, the EC should choose the period most appropriate for the evaluation. EC efforts should focus on the impact and relevance of the NIOSH program in the most recent appropriate period. As a starting point, the ECs might consider three general timeframes:

- 1970-1995, the period from the founding of NIOSH to the initiation of NORA (pre-NORA period)
- 1996-2005 (NORA 1 period)
- After 2005 (NORA 2 period)

Those timeframes are provided as general guidance; the period chosen for review will take into consideration suggestions from the NIOSH research program under review. It is recognized that many of the intermediate and end outcomes documented since 1996 are consequences of research outputs completed before 1996.

III.B.2. Identifying Major Challenges (Figure 2, Circle)

Early in the assessment process, the EC itself should identify the major workplace health and safety challenges for the research program under review. In arriving at a list of challenges, the EC should rely on surveillance findings, including those of NIOSH investigations of sentinel events (through health-hazard or fatality-assessment programs), external advisory inputs, and its own expert judgment. The EC will then be able to compare its own assessment of workplace challenges with the NIOSH program goals and objectives. The congruence between the two will be useful during the assessment of relevance.

III.B.3. Analysis of Research-Program Strategic Goals and Objectives (Figure 2, Box A)

The research program goals and objectives should be evaluated with a focus on how each program goal is related to NIOSH's agency wide strategic goals and to the program challenges identified in the step above (Section III.B.2.). The importance or relevance of an issue may differ from the influence of NIOSH-funded research in addressing it. The EC should recognize that NIOSH research priorities may be circumstantial (for example, congressionally funded) rather than based on NIOSH's assessment of the state of knowledge.

Questions to Guide the Evaluation Committee

1. Are the strategic goals and objectives of the program well defined and clearly described?
2. How well were program goals and objectives aligned with NORA 1 priorities during the last decade?
3. How are current program strategic goals and objectives related to current NIOSH strategy, including NORA 2?
4. Are the research program goals, objectives, and strategies relevant to the major challenges for the research program and likely to address emerging problems in the research program (as determined by the EC while addressing Section III.B.2.)?
   a. Did past program goals and objectives (research and dissemination and transfer activities) focus on the most relevant problems and anticipate the emerging problems in the research program?
   b. Do the current program goals and objectives target the most relevant problems?

Assessment

The EC should provide a qualitative assessment that discusses the relevance of the program's goals, objectives, and strategies in relation to its major challenges.

III.B.4. Review of Inputs (Figure 2, Box B)

Planning inputs include input from stakeholders, surveillance and intervention data, and risk assessments. Production inputs include intramural and extramural funding, staffing, management structure, and physical facilities.

The EC should examine existing intramural and extramural resources and, potentially, prior surveys or case studies that might have been developed specifically to assess progress in reducing workplace illnesses and injuries and to provide information relevant to the targeting of research to future needs. The NIOSH research program should provide the EC all relevant planning and production inputs (see below and Table 2 for examples).

Planning Inputs

Planning inputs can be qualitative or quantitative. Sources of qualitative inputs include the following:

- Federal advisory committees (such as the Board of Scientific Counselors, the Mine Safety and Health Research Advisory Committee, and the National Advisory Committee on Occupational Safety and Health)
- NORA research partners, initial NORA stakeholder meetings, later NORA team efforts (especially strategic research plans), and the NORA Liaison Committee and federal liaison committee recommendations
- Industry, labor, academe, professional associations, industry associations, and the Council of State and Territorial Epidemiologists (CSTE)
- OSHA and MSHA strategic plans and other federal research agendas

Attention should be given to how comprehensive the inputs have been and to what extent gaps in input have been identified and considered by NIOSH.

Sources of quantitative inputs include the following:

- Intramural surveillance information, such as descriptive data on exposures and outcomes (appropriate data may be available from a number of NIOSH divisions and laboratories)
- HHEs
- Reports from the FACE program
- Extramural health-outcome and exposure-assessment data from OSHA, MSHA (both safety and health inspection data), the Bureau of Labor Statistics, the U.S. Department of Defense (DOD), and the U.S. Department of Agriculture (USDA) (fatality, injury, and illness surveillance data); state government partners, including NIOSH-funded state surveillance programs, such as Sentinel Event Notification System of Occupational Risks (SENSOR), Adult Blood Lead Epidemiology and

Surveillance (ABLES), and state-based FACE; and nongovernment organizations, such as the National Safety Council, the Association of Occupational and Environmental Clinics (AOEC), the American Society of Safety Engineers, and the American College of Occupational and Environmental Medicine

- Appropriate data from investigator-initiated extramural research funded by NIOSH

Production Inputs

For the research program under review, NIOSH should identify portions of the NIOSH intramural budget, staff, facilities, and management that play major roles in the research program. Production inputs should be described primarily in terms of intramural research projects, relevant extramural projects (particularly cooperative agreements and contracts), HHEs, and related staff. Consideration should also be given to leveraged funds provided by such partners as the National Institutes of Health (NIH) and the Environmental Protection Agency (EPA) for joint requests for applications or program announcements; and to OSHA, MSHA, and U.S. Department of Defense (DOD) contracts with NIOSH.

Assessment of inputs should include EC consideration of the degree to which allocation of funding and personnel was commensurate with the resources needed to conduct the research and the extent to which funding for the relevant intramural research activity has been limited by lack of discretionary spending beyond salaries (travel, supplies, external laboratory services, and so on). Thus, assessments should consider the adequacy of the qualitative and quantitative planning and production inputs, given the tasks at hand.

Questions to Guide the Evaluation Committee

1. Do planning, production, and other input data promote program goals?
2. How well are major planning, production, and other program inputs used to support the major activities?
3. Is input obtained from stakeholders, including input representing vulnerable working populations and small businesses?
4. Are production inputs (intramural and extramural funding, staffing, management, and physical infrastructure resources) consistent with program goals and objectives?

Assessment

The EC should provide a qualitative assessment that discusses the quality, adequacy, and use of inputs.

III.B.5. Review of Activities (Figure 2, Box C)

Activities are defined as the efforts and work of a program's staff, grantees, and contractors. For present purposes, activities of the NIOSH program under review are divided into research and transfer activities. Table 3 is intended to guide the EC and NIOSH as to the type and organization of information required to evaluate program activities. The table may be incomplete, and some types of research activity may not be applicable to a given NIOSH program. Research activities include safety research, health-outcomes research, safety-design research, and safety-systems research. Transfer activities include information dissemination, training, technical assistance, and education designed to translate research outputs into content and formats that are designed for application in the workplace. Depending on the scope of the program under review, activities may also be grouped by research-program objectives or subprograms.

Conventional occupational safety and health research focuses appropriately on injury, illness, or death; on biomarkers of exposure; and on health effects of new technology, personal protective equipment, and regulations. A focus on surveillance research may be needed when available data inputs are inadequate. A focus on socioeconomic and policy research and on diffusion research is also needed to effect change, because not all relevant intermediate outcomes occur in the workplace. NIOSH may be able to affect important outcomes farther out on the causal chain so as to influence health and safety in the workplace. Other research that might prove important in addressing NIOSH's mission includes the following:

- Surveillance research to assess the degree of significant or systematic underreporting of relevant injuries, illnesses, and biomarkers
- Socioeconomic research on cost-shifting between worker compensation and private insurance
- Research on methods to build health and safety capacity in community health centers that serve low-income or minority-group workers and to improve recognition and treatment of work-related conditions
- Transfer research to change health and safety knowledge of adolescents while they are in high school to improve the likelihood of reduced injuries as they enter the workforce

TABLE 3 Examples of NIOSH Program Research and Transfer Activities

Surveillance (including hazard and injury, illness, and biomarkers of exposure or effect health surveillance and evaluation of surveillance systems)

Health-effects research (illnesses, injuries, and biomarkers):
- Epidemiology
- Toxicology
- Physical and safety risk factors (laboratory-based)
- Development of clinical-screening methods and tools

Exposure-assessment research:
- Chemical hazards
- Physical hazards
- Biologic hazards
- Ergonomic hazards
- Safety (traumatic injury) hazards

Safer-design and safety-systems research

Intervention research:
- Control technologies
    - Engineering controls and alternatives
    - Administrative controls
    - Personal protective equipment
- Work organization
- Community participation
- Policy (such as alternative approaches to targeting inspections)
- Design for safety
- Emergency preparedness and disaster response

Diffusion and dissemination research:
- Training effectiveness
- Information-dissemination effectiveness
- Diffusion of technology

Health-services and other research:
- Access to occupational health care
- Infrastructure—delivery of occupational-health services, including international health and safety
- Socioeconomic consequences of work-related injuries and illnesses
- Worker compensation

Technology-transfer and other transfer activities:
- Information dissemination
- Training programs
- Technical assistance

- Community-based participatory research on differences between recently arrived immigrants and U.S.-born workers regarding perceptions of acceptable health and safety risks so that programs can be targeted to meet the workforce training needs of immigrant workers

Transfer activities should be reviewed to determine whether the NIOSH program appropriately targets its outputs in a manner that will have the greatest impact. Ideally, information dissemination should be proactive, and strategic dissemination should be informed by research on the diffusion of new technologies, processes, and practices. Highly relevant information and technology transfer should include plans for appropriate transfer to all appropriate worker populations, including those considered vulnerable. Training should be incorporated into the strategic goals of all research fields where appropriate.

The EC should review project-level research and transfer activities (including surveillance activities) that have been completed, are in progress, or planned by the program under review. The program under review should provide a list of activities and specify whether they are intramural or extramural. For each extramural project, the key organizations and principal investigators' names should be requested, as should whether the project was in response to a request for proposal or a request for application. For each intramural project, the EC should ask NIOSH to provide a list of key collaborators (from another government agency, academe, industry, or unions).

The EC should evaluate each of the research activities outlined in Table 3 if it forms an important element of the program research. In the case of a sector-based research program (for example, mining or construction) in which health-effects research is not being reviewed, the EC should determine what research inputs influence the program's strategic goals and objective, and then assess the value of the inputs.

Questions to Guide the Evaluation Committee in Assessing Research Activities

1. What are the major subprograms or groupings of activities within the program?
2. Are activities consistent with program goals and objectives?
3. Are research activities relevant to the major challenges of the research program?
   a. Do they address the most serious outcomes?
   b. Do they address the most common outcomes?
   c. Do they address the needs of both sexes, vulnerable working populations, and small businesses?

4. Are research activities appropriately responsive to the input of stakeholders?
5. To what extent are partners involved in the research activities?
6. Are partners involved early in the research process so that they could participate in determining research objectives and research design?
7. Were original resource allocations appropriate for the research activities, and do they remain appropriate?
8. To what extent does peer reviews (internal, external, and midcourse) affect the activities?
9. Is there adequate monitoring of quality-assurance procedures to ensure credible research data, analyses, and conclusions?

Questions to Guide the Evaluation Committee in Assessing Transfer Activities

1. Is there a coherent planned program of transfer activities?
2. Are the program's information dissemination, training, education, technical assistance, or publications successful in reaching the workplace or relevant stakeholders in other settings? How widespread is the response?
3. To what degree have stakeholders responded to NIOSH information and training products?
4. Is there evidence that the formats for information products were selected in response to stakeholder preferences?
5. To what extent do program personnel rely on assessment of stakeholder needs and reactions to prototype information and training projects (formative evaluation techniques)?
6. To what extent does the program build research and education capacity internally and among stakeholders?

Assessment

For this part of the assessment, the EC will provide a qualitative assessment that discusses relevance. This assessment should include consideration of the external factors identified in Section III.A. that constrain choices of research projects and the relevance and effectiveness of transfer activities. The EC should consider the appropriateness of resource allocations. A highly relevant program would address high-priority needs, produce high-quality results, be appropriately collaborative, be of value to stakeholders, and be substantially engaged in transfer activities. A program may be less relevant to the extent that those key elements are not up to the mark or are missing. The discussion should cover those aspects in sufficient detail to arrive at a qualitative assessment of the activities. Assessment of the transfer activities must include considerations of program planning, coherence, and impact. The EC might also consider the incorporation of international research results into NIOSH knowledge-transfer activities for industry sectors in the United States.

III.B.6. Review of Outputs (Figure 2, Box D)

An output is a direct product of a NIOSH research program. Outputs may be designed for researchers, practitioners, intermediaries, and end-users, such as consumers. Outputs can be in the form of publications in peer-reviewed journals, recommendations, reports, Web-site content, workshops and presentations, databases, educational materials, scales and methods, new technologies, patents, technical assistance, and so on. Outputs of the research program's extramurally funded activities should also be considered. Table 4 lists examples of major outputs to be considered by the EC. The NIOSH research program should make every effort to include all pertinent data of the types listed in the table.

Outputs may be tailored to the intended audience to communicate information most effectively and increase the likelihood of comprehension, knowledge, attitude formation, and behavioral intent. The extent of use of formative evaluation data (data gathered before communication for the purpose of improving the likelihood of the intended effects) and the extent of intended user feedback in the design of the output can be considered indicators of appropriate quality assessment.

Some activities such as collaborations can also legitimately be conceptualized as outputs, because the collaboration itself is a result of NIOSH efforts. Cooperation, coordination, more intensive collaboration, and eventual formal partnering can be considered important outputs leading to desirable intermediate outcomes. Technology and knowledge transfer is greatly facilitated through such relationships. The extent of collaboration with other organizations in the determination of research agendas, the conduct of research, the dissemination of research results, and interorganization involvement in the production of outputs can all be measures of output quality and quantity. The EC may consider coauthorship while trying to determine the importance of NIOSH research to the broader research community.

The NIOSH program should provide information on all relevant outputs of the program under review produced during the chosen period.

Questions to Guide the Evaluation Committee

1. What are the major outputs of the research program?
2. Are output levels consistent with resources allocated (were resources allocated and used efficiently to produce outputs)?

TABLE 4 Examples of Research-Program Outputs to Be Considered

Peer-reviewed publications by NIOSH staff:
- Number of original research articles by NIOSH staff
- Number of review articles by NIOSH staff (including best-practices articles)
- Complete citation for each publication
- Complete copies of the "top five" articles
- Collaboration with other public- or private-sector researchers
- Publications in the field of interest with other support by investigators also funded by NIOSH (for example, ergonomic studies with other support by an investigator funded by NIOSH to do ergonomics work, in which case NIOSH should get some credit for seeding interest or drawing people into the field)

Peer-reviewed publications by external researchers funded by NIOSH:
- Number of NIOSH-funded original research articles by external researchers
- Number of NIOSH-funded review articles by external researchers (including best-practices articles)
- Complete citation for each written report
- Complete copies of the "top five" articles
- Collaboration with other government or academic researchers

NIOSH reports in the research program:
- Number of written reports
- Complete citation for each written report
- Complete copies of the "top five" reports

Sponsored conferences and workshops:
- Number of sponsored conferences
- Number of sponsored workshops
- Description of conferences and workshops (title, date, sponsors, target audience, number of participants, and resulting products)
- NIOSH's assessment of value or impact

Databases:
- Number of major databases created by NIOSH staff
- Number of major databases created by external researchers funded by NIOSH grants
- Description of databases:
  - Title, objective (in one to four sentences), and start and stop dates
  - Partial vs. complete sponsorship (if partial, who were cosponsors?)
  - Study or surveillance-system design, study population, and sample size
  - Primary "products" of the database (such as number of peer-reviewed articles and reports)
- Complete copies of the "top two" publications or findings, to date, from each database

continued

TABLE 4 Continued

Recommendations:
- Number of major recommendations
- Description of recommendations:
  - Complete citation (article, report, or conference where recommendation was made)
  - Summary in one to four sentences
  - Percentage of target audience that has adopted recommendation 1, 5, and 10 years later
  - Up to three examples of implementation in the field
- Identification of "top five" recommendations to date

Tools, methods, and technologies (TMT):
- Number of major TMT (includes training and education materials)
- Descriptions of TMT
  - Title and objective of TMT (in one to four sentences)
  - Complete citation (if applicable)
  - Percentage of target audience that has used TMT 1, 5, and 10 years later
  - Up to three examples of implementation in the field
- Identification of "top five" TMT to date

Patents:
- Total number of patents
- For each:
  - Title and objective (in one to four sentences)
  - Complete citation
  - Percentage of target audience that has used product 1, 5, and 10 years later
  - Up to three examples of implementation in the field
- Identification of "top five" patents to date

Miscellaneous:
- Any other important program outputs

3. Does the research program produce outputs that address high-priority areas?
4. To what extent does the program generate important new knowledge or technology?
5. Are there widely cited peer-reviewed publications considered to report "breakthrough" results?
6. What, if any, internal or external capacity-building outputs are documented?
7. Are outputs relevant to both sexes, vulnerable populations, and do they address health disparities?
8. Are outputs relevant to health and safety problems of small businesses?

9. Are products user-friendly with respect to readability, simplicity, and design?
10. To what extent does the program help to build the internal or extramural institutional knowledge base?
11. Does the research produce effective cross-agency, cross-institute, or internal-external collaborations?
12. To what extent does the program build research and education capacity (internal or external)?

Assessment

The EC should provide a qualitative assessment discussing relevance and utility. The outputs of a highly ranked program will address needs in high-priority areas, contain new knowledge or technology that is effectively communicated, contribute to capacity-building inside and outside NIOSH, and be relevant to the pertinent populations. The discussion should cover those aspects in sufficient detail to support the qualitative assessment of the outputs.

III.B.7. Review of Intermediate Outcomes (Figure 2, Box E)

Intermediate outcomes are important indicators of stakeholder response to NIOSH outputs. They reflect the impact of program activities and may lead to the desired end outcome of improved workplace safety and health. Intermediate outcomes include the production by those outside of NIOSH of guidelines or regulations based wholly or partly on NIOSH research (products adopted as national or international public policy or as policy or guidelines by private organizations or industry); contributions to training and education programs sponsored by other organizations; use of publications or other materials by workers, industry, and occupational safety and health professionals in the field; and citations of NIOSH research by industrial and academic scientists.

Intermediate outcomes allow inference that a program's outputs are associated with observed changes in the workplace. Thus, an intermediate outcome reflects an assessment of worth by NIOSH stakeholders (such as managers in industrial firms) about NIOSH research or its products (for example, NIOSH training workshops). Intermediate outcomes that are difficult to monitor but may be valid indicators of relevance or utility include self-report measures by users of NIOSH outputs. Such indicators include the extent to which key intermediaries find value in NIOSH products or databases for the repackaging of health and safety information, the extent to which NIOSH recommendations are in place and attended to in workplaces, and employee or employer knowledge of and adherence to NIOSH-recommended practices.

Questions to Guide the Evaluation Committee

1. Do program outputs result in or contribute to stakeholder training or education activities used in the workplace or in school or apprentice programs? If so, how?
2. Do program activities and outputs result in regulations, public policy, or voluntary standards or guidelines that are transferred to or created by the workplace?
3. Has the program resulted in changes in employer or worker practices associated with the reduction of risk (for example, in the adoption of new feasible control or personal protective technologies or administrative control concepts)?
4. Does the program contribute to changes in health-care practices to improve recognition and management of occupational health conditions?
5. Does the program result in research partnerships with stakeholders that lead to changes in the workplace?
6. To what extent do the program's stakeholders find value in NIOSH products (as shown by document requests, Web-site hits, conference attendance, and so on)?
7. Does the program or a subprogram provide unique staff or laboratory capability that is a necessary national resource? If so, is it adequate, or does it need to be enhanced or reduced?
8. Has the program resulted in interventions that protect both sexes, vulnerable workers, or address the needs of small businesses?
9. To what extent did the program contribute to increased capacity at worksites to identify or respond to safety and health threats?

Assessment

Only a qualitative assessment of product development, usefulness, and impact is required at this point in the EC report. Some thought should be given to the relative value of intermediate outcomes, and the FC recommends applying the well-accepted hierarchy-of-controls model. The discussion could include comments on how widely products have been used or programs implemented. The qualitative discussion should be specific as to the various products developed by the program and the extent of their use by specific entities (industry, labor, government, and so on) for specific purposes. Whether the products have resulted in changes in the workplace or in the reduction of risk should be discussed. The recognition accorded to the program or the facilities by its peers (such as recognition as a "center of excellence" by national and international communities) should be considered in the assessment. To be highly ranked, a program should have high performance in most of the relevant questions in this section. An aspect of the evaluation can be whether the same changes in stakeholder activities and behaviors would probably have occurred without NIOSH efforts.

III.B.8. Review of End Outcomes (Figure 2, Box F)

It is necessary for the EC to assess, to the greatest extent possible, NIOSH's contribution to end outcomes—improvements in workplace health and safety (impact). For purposes of this evaluation, end outcomes are health-related changes that are a result of program activities, including decreases in injuries, illnesses, deaths and exposures or risk. Data on reductions in work-related injuries, illnesses, and hazardous exposures will be available for some programs, and in some cases they will be quantifiable. It is possible, however, to evaluate the impact of a NIOSH research program using either intermediate outcomes or end outcomes. If there is no direct evidence of improvements in health and safety, intermediate outcomes may be used as proxies for end outcomes in assessing impact as long as the EC qualifies its findings. The EC will describe the realized or potential benefits of the NIOSH program. Examples of realized intermediate outcomes are new regulations and widely accepted guidelines, work practices, and procedures, all of which may contribute measurably to enhancing health and safety in the workplace.

The FC recognizes that assessing the causal relationship between NIOSH research and specific occupational health and safety outcomes is a major challenge because NIOSH does not have direct responsibility or authority for implementing its research findings in the workplace. Furthermore, the benefits of NIOSH research program outputs can be realized, potential, or limited to the knowledge gained. Studies that conclude with negative results may nevertheless have incorporated excellent science and contribute to the knowledge base. The generation of important knowledge is a recognized form of outcome in the absence of measurable impacts.

The impact of an outcome depends on the existence of a "receptor" for research results, such as a regulatory agency, a professional organization, an employer, and an employee organization. The EC should consider questions related to the various stages that lead to outputs, such as these:

1. Did NIOSH research identify a gap in protection or a means of reducing risk?
2. Did NIOSH convey that information to potential users in a usable form?
3. Were NIOSH research results (for example, recommendations, technologies) applied?
4. Did the applied results lead to desired outcomes?

Quantitative data are preferable to qualitative, but qualitative analysis may be necessary. Sources of quantitative data include the following:

- Bureau of Labor Statistics (BLS) data on fatal occupational injuries (the Census of Fatal Occupational Injuries) and nonfatal occupational injuries and illnesses (the annual Survey of Occupational Injury and Illnesses)
- NIOSH intramural surveillance systems, such as the National Electronic Injury Surveillance System, the coal-worker x-ray surveillance program, and agricultural-worker surveys conducted by NIOSH in collaboration with USDA
- State-based surveillance systems, such as the NIOSH-funded ABLES, and the SENSOR programs (for asthma, pesticides, silicosis, noise-induced hearing loss, dermatitis, and burns)
- Selected state worker-compensation programs
- Exposure data collected in the OSHA Integrated Management Information System

The FC is unaware of mechanisms for surveillance of many occupationally related chronic illnesses, such as cancers that arise from long exposure to chemicals and other stressors. The incidence and prevalence of many such outcomes are best evaluated by investigator-initiated research. Research that leads to new, effective surveillance concepts or programs warrants special recognition.

The EC should recognize the strengths and weaknesses of outcome data sources. Quantitative accident, injury, illness, and employment data and databases are subject to error and bias and should be used by the EC only for drawing inferences after critical evaluation and examination of available corroborating data. For example, it is widely recognized that occupational illnesses are poorly documented in the BLS Survey of Occupational Injuries and Illnesses, which captures only incident cases among active workers. It is difficult for health practitioners to diagnose work-relatedness of most illnesses that may not be exclusively related to work; furthermore, few practitioners are adequately trained to make such an assessment. Many of those illnesses have long latencies and do not appear until years after people have left the employment in question. Surveillance programs may systematically undercount some categories of workers, such as contingent workers.

In addition to measures of illness and injury, measures of exposure to chemical and physical agents and to safety and ergonomic hazards can be useful. Exposure or probability of exposure can serve as an appropriate proxy for disease or injury when a well-described occupational exposure-health association exists. In such instances, a decrease in exposure can be accepted as evidence that the end outcome of reduced illness or injury is being achieved. That is necessary particularly when the latent period between exposure and disease outcome, as in the case of asbestos exposure and lung cancer, makes effective evaluation of the relevant end outcome infeasible.

As an example of how an exposure level can serve as a proxy, reduction in the number of sites that exceed an OSHA PEL or an American Conference of Governmental Industrial Hygienists threshold limit value is a quantitative measure of improvement of occupational health awareness and reduction of risk. In addition to exposure level, the number of people exposed and the distribution of exposure levels are important. Those data are available from multiple databases and studies of exposure. Apart from air monitoring, such measures of exposure as biohazard controls, reduction in requirements for use of personal protective equipment, and reduction in ergonomic risks are important.

Challenges posed by inadequate or inaccurate measurement systems should not drive programs out of difficult fields of study, and the EC will need to be aware of such a possibility. In particular, contingent and informal working arrangements that place workers at greatest risk are also those on which surveillance information is almost totally lacking, so novel methods for measuring impact may be required.

The commitment of industry, labor, and government to health and safety are critical external factors. Several measures of that commitment can be useful for the EC: monetary commitments, attitude, staffing, and surveys of relative importance. To the extent that resources allocated to safety and health are limiting factors, the EC should explicitly assess NIOSH performance in the context of constraints.

Questions to Guide the Evaluation Committee

1. What are the amounts and qualities of relevant end-outcomes data (such as injuries, illness, exposure, and productivity affected by health)?
2. What are the temporal trends in those data?
3. Is there objective evidence of improvement in occupational safety or health?

4. To what degree is the NIOSH program or subprogram responsible for improvement in occupational safety or health?
5. If there is no time trend in the data, how do findings compare with data from other comparable U.S. groups or the corresponding populations in other countries?
6. What is the evidence that external factors have affected outcomes or outcome measures?
7. Has the program been responsible for outcomes outside the United States that have not been described in another category?

Assessment

The EC should provide a qualitative assessment of the program and subprogram impact, discussing the evidence of reductions in injuries and illnesses or their appropriate proxies.

III.B.9. Review of Potential Outcomes

There may be health and safety impacts not yet appreciated and other beneficial social, economic, and environmental outcomes as a result of NIOSH activities. NIOSH study results may be influential outside the United States, and there may be evidence of implementation of NIOSH recommendations and training programs abroad.

Questions to Guide the Evaluation Committee

1. Is the program likely to produce a favorable change that has not yet occurred or not been appreciated?
2. Has the program been responsible for social, economic, security, or environmental outcomes?
3. Has the program's work had an impact on occupational health and safety in other countries?

Assessment

The EC may discuss other outcomes, including beneficial changes that have not yet occurred; social, economic, security, or environmental outcomes; and the impact that NIOSH has had on international occupational safety and health.

III.B.10. Summary Evaluation Ratings and Rationale

The EC should use its expert judgment to rate the relevance and impact of the overall research program by first summarizing its assessments of the major subprograms and then appropriately weighting the subprograms to determine the overall program ratings.

Table 5 provides some background context to aid the EC in reaching overall ratings for relevance and impact. The EC could consider the items in Table 5 for each subprogram then for the overall program and assess the relevance of the research subprograms and program by reviewing earlier responses to the questions in Sections III.B.2. through III.B.5. (reviews of program challenges, strategic goals and objectives, inputs, and activities). Items 1-4 in Table 5 are pertinent to assessing relevance.

To assess overall impact, the EC first needs to consider the available evidence of changes in work-related risks and adverse effects and external factors related to the changes. The EC should review the responses to the questions in Sections III.B.6. through III.B.8. (reviews of outputs, intermediate outcomes, and end outcomes) and systematically assess the impact of the research program and its subprograms. Items 5-7 in Table 5 will be helpful. The EC should evaluate separately the impact of the research and the impact of transfer activities. Transfer activities occur in two contexts: NIOSH efforts to translate intellectual products into practice and stakeholder efforts to integrate NIOSH results into the workplace. High impact assessments require the EC's judgment that the research program has contributed to outcomes; for example, outcomes have occurred earlier than they would have or are better than they would have been in the absence of the research program,

TABLE 5 Background Context for Program Relevance and Impact

Assess the following for each subprogram:

1. Relevance of current and recently completed research and transfer activities to objective improvements in workplace safety and health.
2. Contributions of NIOSH research and transfer activities to changes in work-related practices and reduction in workplace exposures, illnesses, or injuries.
3. Contributions of NIOSH research and transfer activities to improvements in work-related practices.
4. Contributions of NIOSH research to productivity, security, or environmental quality (beneficial side effects).
5. Evidence of reduction of risk in the workplace (intermediate outcome).
6. Evidence of reduction in workplace exposure, illness, or injuries (end outcome).
7. Evidence of external factors that prevented translation of NIOSH research results into intermediate or end outcomes.

or outcomes would have occurred were it not for external factors beyond NIOSH's control or ability to plan around.

The EC must assign one overall integer score for the *relevance* of the research program to the improvement of occupational safety and health and one overall integer score for the *impact* of the program on the improvement of occupational safety and health. The EC will use its expert judgment, summary assessment of research-program elements, and any appropriate information to arrive at those two scores. In light of substantial differences among the types of research programs that will be reviewed and the challenge to arrive at a summative evaluation of both relevance and impact, the FC chose not to construct an algorithm to produce the two final ratings.

Relevance and impact scores will be based on five-point categorical scales established by the FC (see Boxes 2 and 3) in which 1 is the lowest and 5 the highest rating. The FC has made an effort to establish mutually exclusive rating categories in the scales. When the basis of a rating fits more than one category, the highest applicable score should be assigned. It is up to the EC to determine how individual subprograms should influence final scores. Single integer values should be assigned. Final program ratings will consist of integer scores for relevance and impact and prose justification of the scores.

Box 2 includes the criteria for scoring the overall relevance of the NIOSH research program. As discussed in previous sections, numerous factors can be considered in assessing relevance. The scoring criteria focus on two: the EC assessment

BOX 2
Scoring Criteria for Relevance

5 = Research is in high-priority subject areas and NIOSH is significantly engaged in appropriate transfer activities for completed research projects/reported research results.
4 = Research is in priority subject areas and NIOSH is engaged in appropriate transfer activities for completed research projects/reported research results.
3 = Research is in high priority or priority subject areas, but NIOSH is not engaged in appropriate transfer activities; or research focuses on lesser priorities but NIOSH is engaged in appropriate transfer activities.
2 = Research program is focused on lesser priorities and NIOSH is not engaged in or planning some appropriate transfer activities.
1 = Research program is not focused on priorities and NIOSH is not engaged in transfer activities.

BOX 3
Scoring Criteria for Impact

5 = Research program has made major contribution(s) to worker health and safety on the basis of end outcomes or well-accepted intermediate outcomes.
4 = Research program has made some contributions to end outcomes or well-accepted intermediate outcomes.
3 = Research program activities are ongoing and outputs are produced that are likely to result in improvements in worker health and safety (with explanation of why not rated higher). Well accepted outcomes have not been recorded.
2 = Research program activities are ongoing and outputs are produced that may result in new knowledge or technology, but only limited application is expected. Well accepted outcomes have not been recorded.
1 = Research activities and outputs do not result in or are NOT likely to have any application.

of whether the program appropriately sets priorities among research needs and the EC assessment of how engaged the program is in appropriate transfer activities. Table 6 provides some guidance regarding how the EC may weight research priorities and transfer levels when determining relevance scores.

The EC will consider both completed research and research that is in progress and related to likely future improvements in its assessment of relevance. The EC should keep in mind how well the program has considered the frequency and severity of the problems being addressed; whether appropriate attention has been directed to both sexes, vulnerable populations, or hard-to-reach workplaces; and whether the different needs of large and small businesses have been accounted for. It is up to the EC to determine how to consider external factors in assigning program scores.

Box 3 includes the criteria established for the rating of impact. In general, the EC will consider completed research outputs during the assessment of impact. In assigning a score for impact, it is important to recognize that a "major contribution" (required for a score of 5) does not imply that the NIOSH program was solely responsible for observed improvements in worker health and safety. Many factors may be required to effect improvements. The EC could say that NIOSH made "major contributions" if the improvements would not have occurred when they did without NIOSH efforts.

TABLE 6 Guidance for Weighting Research Priority and Engagement in Appropriate Transfer Activities in the Application of Relevance Score

Assessment of Research Priority	Engagement in Applicable Transfer Activities	Applicable Score
High priority	Significantly engaged	5
High priority	Engaged	4
High priority	Not engaged	3
Priority	Significantly engaged	4
Priority	Engaged	4
Priority	Not engaged	3
Lesser priority	Significantly engaged	3
Lesser priority	Engaged	3
Lesser priority	Not engaged	2
Not focused on priorities	Significantly engaged	2
Not focused on priorities	Engaged	2
Not focused on priorities	Not engaged	1

The FC has some concern that the imposed scoring criteria for impact might be considered a promotion of the conventional occupational-health research paradigm that focuses on health-effects and technology research without much emphasis on the socioeconomic, policy, surveillance, and diffusion research (as opposed to diffusion activities) needed to effect change. The EC should remember that not all intermediate outcomes occur in the workplace. Important outcomes that NIOSH can effect also occur much farther out on the causal chain. NIOSH, for example, has an important role to play in generating knowledge that may contribute to changing norms in the insurance industry, in health-care practice, in public-health practice, and in the community at large. The EC may find that some of those issues need to be addressed and considered as external factors that facilitate or limit application of more traditional research findings. Given the rapidly changing nature of work and the workforce and some of the intractable problems in manufacturing, mining, and some other fields, the EC is encouraged to think beyond the traditional paradigm.

III.C. Assessment of NIOSH Process for Targeting Priority Research Needs and Committee Assessment of Emerging Issues

The second charge to the EC is the assessment of the research program's effectiveness in targeting new research and identifying emerging issues in occupational safety and health most relevant to future improvements in workplace protection. The EC is also asked to provide a qualitative narrative assessment of the program's efforts and to make suggestions about emerging issues that the program should be prepared to address. Among the most challenging aspects of research in illness and injury prevention are the identification of new or emerging needs or trends and the formulation of a research response that appropriately uses scarce resources in anticipation of them.

The EC should review the procedures that NIOSH and the research program have in place to identify needed research relevant to the NIOSH mission and should review the success that NIOSH has had in identifying and addressing research related to emerging issues. It should examine leading indicators from appropriate federal agencies, such as EPA, the Department of Labor, the National Institute of Standards and Technology, NIH, DOD, and the Department of Commerce. Those indicators should track new technologies, new products, new processes, and disease or injury trends.

One source of information deserving particular attention is NIOSH HHE reports. The HHE program offers a potential mechanism for identifying emerging research needs that could be incorporated as input into each of the programs evaluated. The EC should determine whether the program under review appropriately considers pertinent HHE investigation findings. Additional emerging issues may be revealed through consideration of NIOSH and the NIOSH-funded FACE reports, the AOEC reports, the U.S. Chemical Safety Board investigations, and SENSOR and other state-based surveillance programs. Appropriate federal advisory committees and other stakeholder groups should also be consulted to provide qualitative information.

The EC should systematically assess how the research program and its subprograms target new research by evaluating each subprogram for the items listed in Table 7. The EC will have to determine how best to weight subprogram contributions in the program's targeting of new research.

TABLE 7 Targeting of New Research and Identification of Emerging Issues

Assess the following for each subprogram:

1. Past and present effectiveness in targeting most relevant research needs.
2. Effectiveness in targeting research in fields most relevant to future improvements in occupational safety and health.
3. Contribution of NIOSH research to enhancement of capacity in government or other research institutions.

Questions to Guide the Evaluation Committee

1. What information does NIOSH review to identify emerging research needs?
   a. What is the process for review?
   b. How often does the process take place?
   c. How are NIOSH staff scientists and NIOSH leadership engaged?
   d. What is the process for moving from ideas to formal planning and resource allocation?
2. How are stakeholders involved?
   a. What advisory or stakeholder groups are asked to identify emerging research targets?
   b. How often are such groups consulted, and how are suggestions followed up?
3. What new research targets have been identified for future development in the program under evaluation?
   a. How were they identified?
   b. Were lessons that could help to identify other emerging issues learned?
   c. Does the EC agree with the issues identified and selected as important and with the NIOSH response, or were important issues overlooked?
   d. Is there evidence of unwise expenditure of resources on unimportant issues?

The EC members should use their expert judgment both to evaluate the emerging research targets identified by NIOSH and to provide recommendations to NIOSH regarding additional research that NIOSH has not yet identified. Recommendations should include a brief statement of their rationale.

IV. EVALUATION COMMITTEE REPORT TEMPLATE

Consistency and comparability among EC report formats is desirable, but the FC recognizes that each NIOSH research program is different and that each EC is independent. The outline provided in Box 4 flows from the FC's review of NIOSH's generalized logic model (Figure 1), the evaluation flowchart (Figure 2), and the assessment model described earlier in this document. The EC should feel free to use or adapt this outline as necessary when organizing its final report. The FC encourages each EC to look at prior EC reports for organizational ideas.

BOX 4
Suggested Outline for Evaluation Committee Reports

I. Introduction
This section should be a brief descriptive summary of the history of the program (and subprograms) being evaluated with respect to pre-NORA, NORA 1, and current and future plans of the research program presented by NIOSH. It should present the context for the research on safety and health; goals, objectives, and resources; groupings of subprograms; and any other important pertinent information. (A list of the NIOSH materials reviewed should be provided in Appendix C.)

II. Evaluation of Programs and Subprograms (Charge 1)
A. Evaluation summary (should include a brief summary of the evaluation with respect to impact and relevance, scores for impact and relevance, and summary statements).
B. Strategic goals and objectives: should describe assessment of the program and subprograms for relevance.
C. Review of inputs: should describe adequacy of inputs to achieve goals.
D. Review of activities: should describe assessment of the relevance of the activities.
E. Review of research-program outputs: should describe assessment of relevance and potential usefulness of the research program.
F. Review of intermediate outcomes and causal impact: should describe assessment of the intermediate outcomes and the attribution to NIOSH; should include the likely impacts and recent outcomes in the assessment.
G. Review of end outcomes: should describe the end outcomes related to health and safety and provides an assessment of the type and degree of attribution to NIOSH.
H. Review of other outcomes: should discuss health and safety impacts that have not yet occurred; beneficial social, economic, and environmental outcomes; and international dimensions and outcomes.
I. Summary of ratings and rationale.

III. NIOSH Targeting of New Research and Identification of Emerging Issues (Charge 2)
The EC should assess the progress that the NIOSH program has made in targeting new research in occupational safety and health. The EC should assess whether the NIOSH program has identified important emerging issues that appear especially important in terms of relevance to the mission of NIOSH. The EC should respond to NIOSH's perspective and add its own recommendations.

IV. Recommendations for Program Improvement
On the basis of the review and evaluation of the program, the EC may provide recommendations for improving the relevance of the NIOSH research program to health and safety conditions in the workplace and the impact of the research program on health and safety in the workplace.

Appendix A — Framework Document
Appendix B — Methods and Information-Gathering
Appendix C — List of NIOSH and Related Materials Collected in the Process of the Evaluation

B

Materials Provided by the NIOSH Health Hazard Evaluation Program

NIOSH. 2007 (October 2). Evidence for the National Academies' review of the NIOSH Health Hazard Evaluation Program. Produced by the HHE Program staff for the Committee to Review the National Institute for Occupational Safety and Health (NIOSH) Health Hazard Evaluation (HHE) Program. Cincinnati, OH.

NIOSH. 2007 (October 11). HHE Program Stakeholders list provided as suggested stakeholders who could be asked to address the HHE Review Committee. Cincinnati, OH.

NIOSH. 2007 (November 14). NIOSH response #1 to questions for the NIOSH HHE Program from the Committee to Evaluate the NIOSH Health Hazard Evaluation Program. Cincinnati, OH. Included HHE Step Chart, HHE Organizational Chart, 3 Appendices, and 15 numbered and letter reports.

NIOSH. 2007 (November 27). NIOSH response #2 to questions for the NIOSH HHE Program from the Committee to Evaluate the NIOSH Health Hazard Evaluation Program. Cincinnati, OH.

NIOSH. 2007 (December 19 [revised]). NIOSH response #3 to questions for the NIOSH HHE Program from the Committee to Evaluate the NIOSH Health Hazard Evaluation Program. Cincinnati, OH.

NIOSH. 2007 (December 21). NIOSH response #4 to questions for the NIOSH HHE Program from the Committee to Evaluate the NIOSH Health Hazard Evaluation Program. Cincinnati, OH.

NIOSH. 2008 (January 15). NIOSH response #5 to questions for the NIOSH HHE Program from the Committee to Evaluate the NIOSH Health Hazard Evaluation Program. Cincinnati, OH. Included NIOSH HHE decision process (triage flow chart).

NIOSH. 2008 (January 28). NIOSH response to request for more information about the outcomes of the flock HHEs from the Committee to Evaluate the NIOSH Health Hazard Evaluation Program. Included the flock section of the Respiratory Disease Research Program Evidence Package and journal article.

NIOSH. 2008 (February 5). NIOSH response #6 to questions for the NIOSH HHE Program from the Committee to Evaluate the NIOSH Health Hazard Evaluation Program. Cincinnati, OH. Included 8 HHE response letters.

NIOSH. 2008 (February 13). Compendia from the HHE Program. Cincinnati, OH.

NIOSH. 2008 (May 13). HHE Program's Procedures Manual. Cincinnati, OH. Included 42 different attachments.

NIOSH. 2008 (June 3). NIOSH response #7 to questions for the NIOSH HHE Program from the Committee to Evaluate the NIOSH Health Hazard Evaluation Program. Cincinnati, OH.

C

Committee Information Gathering

This appendix provides additional detail regarding the methods used by the National Academies' Committee to Review the National Institute for Occupational Safety and Health (NIOSH) Health Hazards Evaluation (HHE) Program to gather information to carry out its work.

COMMITTEE MEETINGS

As described in Chapter 1, the committee held four meetings during the course of this study. The first three meetings included open sessions for information gathering. The agendas for these open sessions appear in Boxes C-1, C-2, and C-3 below. The fourth meeting was held in closed session.

COMMITTEE REQUESTS FOR INPUT

To receive a broader range of stakeholder input, the evaluation committee requested public input via an online questionnaire (see Chapter 1 for more detailed discussion). The questionnaire was announced via e-mail to over 500 stakeholders, and responses could be submitted online, by e-mail, or by standard mail. The option to respond anonymously was available. A second questionnaire specifically related to HHE Program emergency response activities was also distributed. The questionnaires are included as Boxes C-4 and C-5. A summary of responses can be found in Appendix D.

BOX C-1
Meeting 1, October 18-19, 2007

The National Academy of Sciences Building
2101 Constitution Avenue
Washington, D.C. 20418

OPEN SESSION AGENDA
Thursday, October 18

NIOSH BRIEFINGS: The HHE Program

10:45 a.m. Study Context and Goals, NIOSH Perspective (*Lewis Wade, Ph.D.*, Senior Science Advisor)

11:00 a.m. HHE Overview: Introduction, Strategic Goals, and Evaluation (*Teresa Schnorr, Ph.D.*, Director, Division of Surveillance, Hazard Evaluations, and Field Studies)

11:20 a.m. HHE Overview: Management, Activities, and Outcomes (*Allison Tepper, Ph.D.*, Chief, Hazard Evaluations and Technical Assistance Branch [HETAB])

11:45 a.m. Lunch

12:45 p.m. HHE Program Goal 1: Prevent occupational illnesses through reduced exposure to workplace hazards

- Biological Hazards (*Teresa Seitz, M.P.H., C.I.H.*, Supervisory Industrial Hygiene Team Lead, HETAB)
- Chemical Hazards (*Teresa Seitz*)
- Physical Hazards (*Ken Wallingford, C.I.H.*, Deputy Chief, HETAB)
- Mixed Hazards (*Ken Wallingford*)

1:25 p.m. Question-and-Answer Session: Committee and NIOSH Briefers

1:40 p.m. HHE Program Goal 2: Promote occupational safety and health research on emerging issues (*Kay Kreiss, M.D.*, Chief, Field Studies Branch)

2:00 p.m. Question-and-Answer Session: Committee and NIOSH Briefers

2:15 p.m. Break

2:25 p.m. HHE Program Goal 3: Protect the health and safety of workers during public health emergencies (*Bruce Bernard, M.D., M.P.H.*, Supervisory Medical Team Lead, HETAB)

2:45 p.m. Question-and-Answer Session: Committee and NIOSH Briefers

3:00 p.m. HHE Program Future Directions (*Allison Tepper*)

3:10 p.m. Closing Question-and-Answer Session: Committee and NIOSH Briefers

3:30 p.m. **PUBLIC COMMENT SESSION**

4:00 p.m. End of Open Session

BOX C-2
Meeting 2, December 10-11, 2007

The Keck Center of the National Academies
500 Fifth Street, NW
Washington, D.C. 20001

OPEN SESSION AGENDA
Monday, December 10

10:15 a.m. Public Welcome and Brief Committee Member Introductions (*Rogene Henderson*, Chair)

10:20 a.m. Committee's task and importance of stakeholder input

STAKEHOLDER PANELS

Each session will include a question-and-answer session among committee and discussants

10:30 a.m. **Workplace Sector Panel**
Moderator: Rosemary Sokas, committee member

Discussants:
Shelley Davis, Farmworker Justice, Washington, DC
Janie Gittleman, Center to Protect Workers' Rights, Silver Spring, MD
José Olíva, Interfaith Worker Justice, Chicago, IL
Frank Renshaw, Rohm and Haas Company, Corydon, PA

11:45 a.m. Lunch

1:00 p.m. **Public Sector Panel**
Moderator: Barbara Silverstein, committee member

Discussants:
Marthe Kent, OSHA Region 1, Boston, Massachusetts
Andrea Kidd-Taylor, Community Health and Policy, Morgan State University, Baltimore
Kenneth Rosenman, Occupational and Environmental Medicine, Michigan State University, East Lansing
Joshua Sharfstein, Baltimore City Health Department

2:15 p.m. Break

DISCUSSION WITH NIOSH

2:30 p.m. Discussion and clarification of NIOSH responses to committee questions posed after the first meeting

4:00 p.m. End of open session

BOX C-3
Meeting 3, January 15-16, 2008

Arnold and Mabel Beckman Center of the National Academies, Board Room
100 Academy Drive Irvine, CA 92617

OPEN SESSION AGENDA
Tuesday, January 15

10:15 a.m. Public Welcome and Brief Committee Member Introductions (*Rogene Henderson*, Committee Chair)

10:20 a.m. Committee's task and importance of stakeholder input

10:30 a.m. Input regarding HHE conducted for the Alameda County Public Authority for In-Home Services (HETA 2001-0139-2930)

Linda Ayala, Training and Outreach Coordinator, Public Authority for In-Home supportive Services in Alameda County, CA (via telecom)

11:00 a.m. Input regarding HHEs conducted at Kaiser Permanente (HETA 2003-0280-2974)

Peggy Hoffman, Kaiser Permanente, Redwood City, CA
Barbara Smisko, Kaiser Permanente, Santa Teresa, CA (via telecom)

11:30 a.m. Input regarding HHE conducted at U.S. Magnesium (HETA 2004-0169-2982)

Tom Tripp, Technical Services Coordinator, U.S. Magnesium, Rowley, Utah (via telecom)
Bryant Hardy, Union Representative, U.S. Magnesium, Rowley, Utah
Michael Wright, Director, Health Safety and Environment, United Steelworkers of America

12:00 p.m. Lunch

1:00 p.m. *Barbara Materna, Ph.D.*, Chief, Occupational Health Branch, California Department of Public Health

1:30 p.m. Input regarding HHE conducted at Carmi Flavor and Fragrance Company (HETA 2006-0303-3043)

Roger Speakman, VP Manufacturing, Carmi Flavor and Fragrance Company
Barbara Materna, Ph.D., Chief, Occupational Health Branch, California Department of Public Health

2:00 p.m. Opportunity for Comments from NIOSH

2:30 p.m. End open session

BOX C-4
Text of Broadcast Questionnaire

Opportunity for Input to the National Academies Committee to Review the NIOSH Health Hazard Evaluation Program

Dear Colleague:

The Committee to Review the National Institute for Occupational Safety and Health (NIOSH) Health Hazard Evaluation (HHE) Program was formed by the National Research Council to review the impact, relevance, and future directions of the program. The study is being conducted under the auspices of the Division of Earth and Life Studies. The committee's first meeting was held October 18-19, 2007. Your input would be valuable for consideration at future meetings on December 10-11, 2007, and January 15-16, 2008.

In addition to evaluating what the HHE Program produces, the committee will determine whether it is appropriate to credit the program with changes in workplace practices, hazardous exposures, and/or occupational illnesses and injuries, or whether the changes are the result of other unrelated factors. In conducting its assessment, the committee will evaluate

A. The impact of the HHE Program on

- Reducing worker risk and preventing occupational illness in investigated workplaces;
- Transferring program-generated information to relevant employers and employees beyond the investigated workplaces;
- NIOSH research and policy-development programs; and
- The activities of regulatory agencies, occupational safety and health professionals and organizations, state and local health agencies, and others in the occupational health community, as achieved by transferring program-generated hazard and prevention information.

B. The relevance of the HHE Program in addressing current and emerging workplace health hazards.

The committee has generated a list of questions (see below) to guide your input.

QUESTIONS

1. Are you familiar with the NIOSH Health Hazard Evaluation (HHE) Program activities and reports related to occupational safety and health?

2. How did you find out about the HHE Program (e.g., website, a recommendation)?

BOX C-4
Continued

3. If you have made an HHE request, what kind of experience have you had? Please be explicit in your response.

4. Have you ever considered requesting an HHE and ultimately decided against making the request? If so, why?

5. Do you have any suggestions about improving access to information about the program?

6. What are the most serious occupational health hazards you have encountered in your work, and how could the HHE Program help you address these issues?

7. What do you see as significant emerging health hazards in occupational safety and health?

8. What is your occupation or to what type of organization or industry to you belong? Please feel free to provide your name (optional). Any information you provide will help the committee understand the context for your responses.

BOX C-5
Text of Emergency Response Questionnaire

Dear Colleague:

The Committee to Review the National Institute for Occupational Safety and Health (NIOSH) Health Hazard Evaluation (HHE) Program was formed by the National Research Council to review the impact, relevance, and future directions of the program. The study is being conducted under the auspices of the Division of Earth and Life Studies. The committee has held meetings October 18-19, 2007, December 10-11, 2007, and January 15-16, 2008. The committee's final meeting will be held February 21-22, 2008. The committee requests your input as a stakeholder of the HHE Program's emergency response activities.

In addition to evaluating what the HHE Program produces, the committee will determine whether it is appropriate to credit the program with changes in workplace practices, hazardous exposures, and/or occupational illnesses and injuries, or whether the changes are the result of other unrelated factors. In conducting its assessment, the committee will evaluate

A. The impact of the HHE Program on

- reducing worker risk and preventing occupational illness in investigated workplaces;
- transferring program-generated information to relevant employers and employees beyond the investigated workplaces;
- NIOSH research and policy-development programs; and
- the activities of regulatory agencies, occupational safety and health professionals and organizations, state and local health agencies, and others in the occupational health community, as achieved by transferring program-generated hazard and prevention information.

B. The relevance of the HHE Program in addressing current and emerging workplace health hazards.

You can learn more about the committee and its charge at the following website: http://www8.nationalacademies.org/cp/projectview.aspx?key=48818

The committee has generated a list of questions (see below) to guide your input. If you wish to contribute, please respond via email to me at your earliest convenience. Keep in mind that any written comments submitted to the committee are required to be included in the study's public access file. If you wish your comments to remain anonymous, the committee has developed an on-line questionnaire in which providing your name and affiliation is optional (http://www8.nationalacademies.org/survey/dels/nioshhhe.htm). The

BOX C-5
Continued

questions on the website, however, were not specifically designed to learn about HHE Program emergency response activities as the questions below are.

Feel free to respond as you see most appropriate and to forward this request for input to other professionals.

Questions regarding NIOSH HHE Program emergency response activities

1. What was the nature of your experience with NIOSH HHE emergency response staff who worked with your agency?

2. What was the NIOSH HHE staff role in the investigation or response? Was it relevant and/or appropriate in your opinion? How could it have been changed? What would you suggest doing in the future if similar circumstance arose?

3. What do you feel was the overall quality of work performed by the HHE staff responders? Was their arrival timely? Were reports prepared in a timely fashion?

4. What was the overall impact of the NIOSH HHE staff involvement in the investigation or emergency response? Please address specifically the HHE Program's impact on protection of workers from health hazards during the incident/investigation. To what extent were they successful? Were there any measurable changes in process/procedures/worker protection as a result of HHE Program staff involvement?

5. What is your occupation or to what type of organization or industry do you belong? Please feel free to provide your name (optional). Any information you provide will help the committee understand the context for your responses.

D

Stakeholder Input: Key Recommendations and Emerging Health Hazards

RECOMMENDATIONS FROM STAKEHOLDERS REGARDING THE HHE PROGRAM

In response to requests for input via an online questionnaire (see Chapter 1 and Appendix C) and during committee meeting discussions, stakeholders made useful comments and suggestions regarding the Health Hazard Evaluation (HHE) Program. The comments are summarized in this appendix. A table of original responses is available through the Public Access Records Office of the National Academies (Stakeholder Response Table, 2008).

Comments were received from occupational, environmental, and internal physicians, medical directors, and nurses; environmental safety specialists and administrators; industrial hygienists; toxicologists; epidemiologists; research psychologists; engineers; academics; compliance officers; investigators; nonprofit health organization representatives; union and worker representatives; and federal and state representatives. Some respondents have worked with the National Institute for Occupational Safety and Health (NIOSH) or the Occupational Safety and Health Administration (OSHA). Some have had direct experience with HHE investigations.

Comments and recommendations related to program improvement are divided into the following categories: strategic planning, interactions with the occupational health community, surveillance, conduct of evaluations, evaluation reports, product dissemination, and training. Comments related to emerging issues are summarized in the final section of this appendix. It should be noted that the comments summarized here do not necessarily reflect the views or opinions of the committee, although they may have influenced the deliberations that led to the committee's conclusions.

HHE Program Improvements

Strategic Planning

- Improve the connection between NIOSH management and the front-line science staff to allow the program to become more science oriented;
- Prioritize hazards to be studied, balancing larger issues with other issues needing to be addressed;
- Conduct targeted investigations within certain industries, such as poultry and food service; and
- Establish rapid response teams to encourage timeliness in response to HHE requests.

Interaction with the Occupational Health Community

- Communicate better with safety professionals in industry;
- Market the value of HHEs through collaborations with relevant professional associations and employer or business groups (e.g., National Business Group on Health);
- Partner with community institutions, worker advocate groups, faith-based centers, and day labor organizations to relate to immigrant workers;
- Improve the visibility of the HHE Program to employers, unions, occupational medicine physicians and nurses, health departments, and workers to make the program more accessible;
- Encourage state-based programs to make referrals for HHEs;
- To assist local health department occupational health professionals, consult with local experts in nearby field offices in order to clarify procedures related to the advisability of referrals to the HHE Program (Determining which of the federal, state, and local agencies responsible for responding to specific issues is difficult);
- Work with OSHA to communicate particularly with small- and midsized organizations and encourage employers with an evaluation program to use the OSHA consultation program;
- Develop partnerships with the Centers for Disease Control and Prevention and the Environmental Protection Agency to collaborate with the federally designated pediatric environmental health specialty units to conduct HHEs in day cares and schools;

- Improve adherence to notification protocol for informing local public health agencies of HHEs being conducted in their jurisdictions. Such communication could result in access to relevant historical and recent testing data;
- Encourage the use of pollution prevention or toxic use reduction strategies to reduce hazards at the source, and seek the advice of other public technical assistance agencies that might be able to help;
- Ensure that OSHA compliance officers are informed about the HHE Program and how it can be accessed by employers;
- Market the HHE Program and reports to faculty and students of NIOSH-funded Education and Research Centers;
- Reach out to state and local public health officers and OSHA programs to promote better public health infrastructure;
- Send reports to state and local government agencies and to for-profit and nonprofit organizations;
- Promote the HHE Program to unions, workers (organized and unorganized), businesses, nongovernmental organizations, occupational and environmental health professionals, healthcare providers, and state and local health departments;
- Target the construction sector because of its limited resources and high risks in both safety and health issues; and
- Educate healthcare facilities about exposures in the hospital environment and collaborate with NIOSH researchers to follow up on issues identified through HHEs.

Surveillance

- Create a national surveillance system for occupational illness;
- Strengthen ties between NIOSH and state health departments (e.g., funding, cooperative agreements) to create surveillance programs in as many states as possible;
- Look to workers' compensation for trend data;
- Collect and study the trend data to determine sentinel events;
- Work with the U.S. Chemical Safety and Hazard Investigation Board to address chemical hazards; and
- Actively solicit reports of unusual illnesses or patterns of illnesses among workers from occupational physicians, companies, and workers.

Conduct of Evaluations

- Make suggestions at the time of worksite investigations, initiate more direct telephone contact, and provide interim reports to ensure rapid implementation of recommended hazard control solutions;
- Have more sophisticated neuropsychological testing of workers exposed to neurotoxic metals, solvent, and pesticides; and
- Follow up with individual workplaces and requestors regarding the feasibility and implementation of recommendations on an ongoing basis.

Evaluation Reports

- Improve the short summaries of completed work products;
- Prepare both short and full reports;
- Prepare one-page summaries of reports in layperson language for more wide distribution of the reports;
- Stress with an additional paragraph in each HHE report that the causes of all injuries (e.g., fatigue, work organization, equipment failures, training) should be identified and solutions implemented to prevent their recurrence; and
- Prioritize HHE recommendations by effectiveness and provide a prioritized listing of available literature to assist OSHA in more specific rule-making.

Product Dissemination

Online Dissemination

- Improve the online search engine, including provision of an alphabetical categorization by main topic (exposure or disease);
- Increase links of HHE reports to OSHA and NIOSH webpages;
- Post HHE Program success stories on the OSHA website;
- Post online fact sheets and other materials for workers in languages other than English;
- Provide online content to elicit more commentaries from safety and health practitioners; and
- Provide subscriptions to selective e-mail HHE-specific brief reports (e.g., NIOSH eNews).

Other Dissemination

- Make information available to all stakeholders, including employers and workers including websites and pamphlets for occupational safety and health departments to stimulate further exploration of specific problems;
- Provide public information that distinguishes NIOSH from Immigration and Customs Enforcement and other government agencies;
- Post information in all workplaces;
- Bring more attention to publication of sentinel new findings, case reports, analyses of trends, and so forth, in peer-reviewed professional journals, including the publication of summaries of HHE reports of particular interest;
- Reference and link to HHEs as "one-liners" in *Morbidity and Mortality Weekly Report* in a timely manner to provide immediate access to information;
- Publicize more, including dissemination through trade organizations and journals, trade and union fairs, and conferences;
- Advertise through better business bureaus, wholesale marketers, day labor organizations, and Telemundo to reach immigrant populations; and
- Advertise in local newspapers.

Training

- Provide field placement training in NIOSH for students and inform professors about the HHE Program to help promote the field of occupational health; and
- Provide training to OSHA staff about the HHE Program.

Emerging Health Hazards

Stakeholders identified emerging health hazards with broad comments about a number of wide-ranging issues and concerns, as follows:

- All-hazards disaster preparedness, emergency response, and terrorist-related defense or assessments;
- Unforeseen issues associated with climate change, water shortages, and other environmental extremes;
- Continuing development of new compounds and chemicals, including the use of chemicals about which there are limited or no health data;
- The downsizing of health and safety staffs across the country and weak enforcement of safety and health laws;
- Emerging issues related to the aging workforce;

- The identification of broad causes of injury in each plant and industry;
- The effect of abuse by supervisors and work stress on the reporting of problems in the workplace;
- The organization of work and resulting stress, including stresses caused by workload, shifts, shift work, and their consequences;
- Unrecognized or uncharacterized respiratory hazards related to work organization, job stress, psychosocial disorders, and musculoskeletal disorders;
- Cultural and civility issues, expectations such as inappropriate public behaviors or workplace violence; and
- Communication.

Specific hazardous exposures and health effects were cited by stakeholders, including

- impulse noise;
- serum chromium or beryllium among welders with metal on metal hip implants;
- welding fumes;
- lead;
- respiratory disease and exposures in many industries;
- work-related asthma;
- silica and coal dusts;
- fiberglass dust;
- sugar dust;
- new technologies in nanotechnology, pharmaceuticals, and biotechnology;
- bioaerosol exposure and organic dust;
- radiation;
- surgical smoke plumes from all types of sources;
- chemical and hazardous drug exposures (including exposures associated with chemotherapy);
- solvents;
- isocyanates and aziridine (for example, as used in golf ball manufacturing and rubber coating);
- cadmium;
- pesticide application;
- neurotoxicants and their effect on pregnant or breastfeeding women;
- unregulated or under-regulated toxins identified as carcinogens;
- multiple chemical exposures during and after hardwood floor finishing;

- indoor environmental quality issues, including water-damaged buildings, mold growth, and poor air quality;
- mycotoxins and endotoxins;
- bacteria (e.g., methicillin-resistant *Staphylococcus aureus*) and emerging infectious diseases;
- formaldehyde in particle board furniture;
- mixed low-level exposures;
- long-standing hazards not covered by OSHA standards including cleaning products (e.g., disinfectants), hexavalent chromium in portland cement and concrete, and diesel exhaust;
- carbon monoxide exposure (e.g., from gasoline generators or gas-powered washers);
- establishment of occupational exposure limits;
- dermatitis;
- effects of food additives;
- behavioral modification for the protection against needlesticks and other contaminated sharps injuries in hospitals;
- hazards associated with construction and warehousing;
- agricultural hazards;
- sources of autoimmune diseases in women and healthcare provider immune system problems; and
- musculoskeletal issues (e.g., ergonomic issues related to text messaging and small-computer use or associated with multiple operators of the same piece of equipment).

E

Biographical Sketches

COMMITTEE MEMBERS

ROGENE F. HENDERSON, Ph.D., (*Chair*) is a senior biochemist and toxicologist emeritus in the Experimental Toxicology Program of the Lovelace Respiratory Research Institute, and a clinical professor in the College of Pharmacy at the University of New Mexico in Albuquerque. Her major research interests are in the use of bronchoalveolar lavage fluid analyses to detect and characterize biomarkers of developing lung disease, the toxicokinetics of inhaled vapors and gases, and the use of biological markers of exposure and of effects to link environmental exposure to disease. She has served on a number of Department of Energy, Environmental Protection Agency, and National Institute of Environmental Health Sciences, and U.S. Army scientific advisory boards, and was recently appointed chair of EPA's Clean Air Scientific Advisory Committee. Dr. Henderson is a national associate of the National Academies who has chaired and served as a member of a number of committees, including as a member of the Board on Environmental Studies and Toxicology. She received her Ph.D. in chemistry from the University of Texas.

JOEL BENDER, M.D., Ph.D., has been corporate medical director for General Motors (GM) Health Services since May 2002 where he directs GM's global occupational health programs and policies. He also acts as a healthcare consultant for GM, directs work life programs for salaried employees, serves as a government liaison with health-related agencies, and is responsible for health promotion activities for

GM employees and their family members. Prior to joining GM, Dr. Bender worked in the field of occupational and environmental medicine for nearly 25 years, in leadership positions with the Campbell Soup Company, Owens-Corning, and Dupont. His activities have included clinical, executive, governing, and medical advisory roles in many national, international, and professional organizations, and he recently served a four-year term on the Centers for Disease Control and Prevention's Advisory Committee. He received an M.D. from the University of Alabama's School of Medicine, a Ph.D. in environmental sciences from Drexel University, and is certified as a specialist in occupational and environmental medicine.

EULA BINGHAM (IOM), Ph.D., is professor of environmental health at the University of Cincinnati College of Medicine. Her interests include risk assessment, regulatory toxicology, environmental carcinogenesis, and occupational health surveillance. She was assistant secretary for labor for the Occupational Safety and Health Administration (OSHA) from 1977-1981. Throughout her career, Dr. Bingham has served on numerous national and international advisory groups, including advisory committees of the Food and Drug Administration, Department of Labor, National Institute for Occupational Safety and Health, National Institutes of Health, Natural Resources Defense Council, and the International Agency for Research on Cancer. The committees addressed issues concerning research needs in health risk assessment and the potential health effects of environmental exposure to chemicals. Dr. Bingham has a Ph.D. from the University of Cincinnati in zoology (physiology), ecology, and biochemistry. She is an IOM member who has served on a number of committees of the National Academies.

JAMES E. CONE, M.D., M.P.H., is the medical director of the World Trade Center Health Registry, in the Division of Epidemiology at the New York City Department of Health and Mental Hygiene. He previously directed programs in environmental and occupational medicine at the city and state levels, and in the university environment, including over 20 years of experience teaching, performing clinical and applied epidemiological research, and practicing clinical occupational and internal medicine. Dr. Cone is the immediate past chairperson of the Occupational Health and Safety Section of the American Public Health Association. His research interests are in occupational and environmental health, particularly toxic effects of exposures in the indoor environment, and health and mental health consequences of chemical and other environmental disasters, spills, and explosions, fires, and building collapses. He holds an M.D. from the University of California, San Francisco and an M.P.H. from the University of California, Berkeley. Other training in occupational and internal medicine includes an internship and residencies at the Cook County Hospital in Chicago and the Worcester Memorial Hospital in Massachusetts, as well as a residency in epidemic intelligence service training/occupational medicine in the Hazard Evaluation System and Information Service Branch of the Division of Surveillance, Hazard Evaluations and Field Studies at the Centers for Disease Control and Prevention, National Institute for Occupational Safety and Health in Cincinnati, Ohio.

MONICA GAUGHAN, Ph.D., M.P.A., is an assistant professor of health policy and management in the University of Georgia's College of Public Health. Her research interests focus on scientific and technical careers, research evaluation, science policy, health disparities, and reproductive health. Her research is currently supported by an early CAREER award from the National Science Foundation. She has been a faculty member at Oglethorpe University, where she served as director of the Rich Foundation Urban Leadership Program, and the Georgia Institute of Technology. She served as a presidential management intern in the U.S. Department of Health and Human Services from 1992 to 1994, where she had assignments in the Offices of the Assistant Secretary for Health, the Assistant Secretary for Legislation, the Health Resources and Services Administration, and the National Institutes of Health. She also has experience working in the community mental health system as a case manager and substance abuse counselor. Dr. Gaughan earned her Ph.D. in sociology from the University of North Carolina at Chapel Hill and her M.P.A. from the Maxwell School of Citizenship and Public Affairs, Syracuse University.

CLARION JOHNSON, M.D., is currently the global medical director, Medicine and Occupational Health, for Exxon Mobil Corporation, which delivers industrial hygiene, drug testing, and health promotion services to over 80,000 ExxonMobil and affiliate employees who are engaged in exploration and production in a number of challenging environments worldwide. Dr. Johnson received his M.D. from Yale University School of Medicine and is board certified in internal medicine, cardiology, and occupational medicine. In addition to a cardiology fellowship, Dr. Johnson did a military/basic science fellowship at Walter Reed Army Institute of Research, followed by two years as a postdoctoral student in the field of microwave research. He is currently a board member of the Milbank Memorial Fund and National Business Group on Health, and is the former chairman of the Virginia Health Care Foundation. He has published a variety of articles in various fields.

FRANKLIN E. MIRER, Ph.D., has recently assumed a faculty position at the Department of Environmental and Occupational Health Sciences at Hunter College after 25 years as Director of the Health and Safety Department for the United Automobile, Aerospace and Agricultural Implement Workers of America (UAW). His primary scientific interest is exposure and risk assessment in the occupational environment. Dr. Mirer currently serves on the National Academies Committee to Review NIOSH Research Programs, and in the past served on committees on Institutional Means for Risk Assessment, Risk Assessment Methodology, and the Review of the Health Effects Institute. He has testified before House and Senate committees on occupational safety and health matters and before OSHA on many health and safety standards. Dr. Mirer was inducted into the National Safety Council's Health and Safety Hall of Fame and is a fellow of the Collegium Ramazzini and the American Industrial Hygiene Association. He has a Ph.D. in physical organic chemistry from Harvard University and is a toxicologist and certified industrial hygienist.

BARBARA SILVERSTEIN, Ph.D., M.P.H., is the research director of the Washington State Department of Labor & Industries Safety and Health Assessment and Research for Prevention (SHARP) Program. She has worked on ergonomics-related issues at OSHA, the University of Michigan Center for Ergonomics, the Finnish Institute of Occupational Health, and the California Department of Health Services. Her major areas of research have been identification and control of work-related musculoskeletal disorders and comparison of surveillance methods and intervention studies to control these disorders. She has conducted field research in a number of industries, including electronics, meat, poultry and fish processing, newspaper publishing, appliance manufacturing, medical equipment manufacturing, office work environment, pulp and paper mills, aluminum mills, saw mills, construction, automotive manufacturing, and nursing homes. She serves on a number of national and international commissions and editorial boards regarding occupational safety and health. She received her M.S. from the University of California, San Francisco, her M.P.H. in epidemiology and environmental and industrial health from the University of Michigan, and her Ph.D. in epidemiologic science from the University of Michigan.

ROSEMARY K. SOKAS, M.D., M.Sc., M.O.H., is professor of environmental and occupational health sciences at the University of Illinois at the Chicago School of Public Health, and director of the Illinois Public Health Research Fellowship Program. Her research interests include applied intervention effectiveness studies targeting occupational safety and health needs of vulnerable working populations. She previously served on the faculties of the University of Pennsylvania School of Medicine and George Washington University, directed OSHA's Office of Occupational Medicine, and served as associate director for science at NIOSH. Dr. Sokas is currently a member of the National Academies' Committee on the Review of NIOSH Research Programs, and has served on the Committee on Persian Gulf Syndrome Comprehensive Clinical Evaluation Program and the Committee to Review the Worker and Public Health Activities Program administered by the Department of Energy and the Department of Health and Human Services. She has an M.D. from the Boston University School of Medicine and an M.Sc. and an M.O.H. from the Harvard School of Public Health (occupational physiology and occupational health, respectively).

MICHAEL J. WRIGHT, M.S., is the director of Health, Safety and Environment for the United Steelworkers, a labor union representing American and Canadian workers in the steel, paper, oil, chemical, rubber, forestry, mining, and other industries. He has held this position since 1983. His areas of expertise include risk communication, occupational and environmental health, global health and safety, and industrial hygiene. Mr. Wright has considerable experience with the National Institute for Occupational Safety and Health's Health Hazard Evaluation Program, having requested and participated in numerous health hazard evaluations on behalf of workers he represents. He was a member of the National Academies Committee on Industrial Competitiveness and Environmental Protection and the National Advisory Committee on Occupational Safety and Health. Mr. Wright holds an M.S. degree from the Harvard School of Public Health.

NRC STAFF

SAMMANTHA L. MAGSINO is a program officer with the National Research Council's Board on Earth Sciences and Resources. She was previously a geologist with the Washington State geologic survey, where she produced earthquake hazard maps and served as a technical advisor to the state on volcanic hazards. Previously, she was the science coordinator for a National Science Foundation facility at The University of Texas at Austin conducting aerogeophysical surveys in Antarctica, and has worked for the Center for Nuclear Waste Regulatory Analyses at the Southwest Research Institute conducting geophysical investigations near the proposed Yucca Mountain Nuclear Waste Facility in support of volcanic hazard assessment. She holds an M.S. degree in geology from Florida International University.

SUSAN R. MCCUTCHEN is a senior program associate with the Institute of Medicine's Board on Health Sciences Policy. She has been on staff at the National Academies for 27 years and has worked in several institutional divisions and with many different boards, committees, and panels within those units. The studies in which she has participated have addressed a broad range of subjects and focused on a variety of issues related to science and technology for international development, technology transfer, aeronautics and the U.S. space program, natural disaster mitigation, U.S. education policy and science curricula, needle exchange for the prevention of HIV transmission, the scientific merit of the polygraph, human factors/engineering, research ethics, disability compensation programs, health hazard evaluation, and medical and public health preparedness for catastrophic events. Ms. McCutchen has a B.A. in French, with minors in Italian and Spanish, from Ohio's Miami University, and an M.A. in French, with a minor in English, from Kent State University.

TONYA E. FONG YEE is a senior program assistant with the National Research Council's Board on Earth Science and Resources. Before coming to the National Academies, she interned at the Substance Abuse and Mental Health Services Administration, working on the Safe Schools/Healthy Students Initiative. She received her B.S. in business administration, with a focus on marketing, from the University of Florida in Gainesville, Florida.